Water in Synthetic Fuel Production

Water in Synthetic Fuel Production

The Technology
and Alternatives

Ronald F. Probstein
Water Purification Associates
and
Massachusetts Institute of Technology

Harris Gold
Water Purification Associates

The MIT Press
Cambridge, Massachusetts, and London, England

Any opinions, findings, conclusions or recommendations herein are those of the authors and do not necessarily reflect the views of the National Science Foundation.

This book was set in VIP Times Roman by the Composing Room of Michigan, Inc. It was printed and bound by Halliday Lithograph Corporation in the United States of America.

Library of Congress Cataloging in Publication Data

Probstein, Ronald F
  Water in synthetic fuel production.

  Includes bibliographies and index.
  1. Synthetic fuel industry—Water supply.
2. Synthetic fuel industry—Waste disposal. I. Gold,
Harris, joint author. II. Title.
TP360.P76   662'.66   78-1815
ISBN 0–262–16071–4 (hardcover)
ISBN 0–262–66039–3 (paperback)

This book is dedicated to the idea that America will choose wisely to ensure the continuing availability of energy supplies.

Contents

Preface

Plants to manufacture synthetic fuels from coal and oil shale require large quantities of fresh water and produce large quantities of dirty water. In the United States this poses a problem; much of the easily mined coal and almost all of the high grade oil shale are in the arid West, and local and temporal water shortages sometimes occur where coal supplies are located in the East. In all regions the discharge of contaminated water is constrained by environmental considerations. In this book we have endeavored to present the practically achievable technology that can be incorporated in synthetic fuel plants to minimize water consumption and pollution. The book is intended to be a guide to understanding the role water plays in synthetic fuel production and includes the basic concepts underlying water usage and water treatment in this context.

Not all of the information needed for our analyses was known with certainty at the time of writing, and many estimates given will undoubtedly have to be revised. Moreover, the technologies to produce synthetic fuels should be expected to change. However, by concentrating on the fundamental principles of coal and oil shale conversion as they relate to water needs, the methodologies and results should be generally valid. Our aim was to show what can be done to minimize the water-related environmental impacts of commercial size plants before they are built and not afterward.

Although the book does try to present a reasonably comprehensive picture of synthetic fuel production, it is not a design manual nor does it attempt to examine every possible process or product fuel. For example, the conversion of tar sands is not discussed at all, while only brief note is made of underground (*in situ*) conversion technologies and physical and chemical coal cleaning.

The book is directed to a wide audience including those responsible for planning energy development, those involved with the engineering and design of synthetic fuel plants, and students and others who desire a background in synthetic fuel production. The book is formally self-contained and all the material—encompassing the disciplines of chemical, mechanical, civil, environmental, and mining engineering, water resources, and chemistry—should be accessible to anyone with an undergraduate degree in engineering or the physical sciences. For those readers without a technical background Chapters 1 to 3 and the concluding Chapter 9 can be read with little difficulty. Chapters 4 to 8 have their main results and conclusions summarized at the end of each chapter.

Through directly cited references, every effort has been made to acknowledge the work of others discussed in the book. However, the book also contains much unpublished work of the authors. Some of this work has been taken from ongoing studies being carried out at Water Purification Associates by the authors and their colleagues.

The support of this project is an indirect expression of concern and interest in the development of a sound policy on the part of the United States toward the production of synthetic fuels as an alternative energy source. It is to this idea that the book is dedicated.

Cambridge, Massachusetts   Ronald F. Probstein
October, 1977   Harris Gold

Acknowledgments

We are deeply grateful to the National Science Foundation, which through its former program of Research Applied to National Needs sponsored the preparation of this book before the importance of the subject matter was widely appreciated. Special thanks are due Edward H. Bryan who served as the program manager of the initial grant under which the work was carried out. Much of the credit for the completion of the book goes to Richard H. Warder, Jr., who became the program manager on the continuation grants. Without his support the program might never have gone to completion because of continually altering government priorities. Donald Senich and Harold A. Spuhler shepherded the project through its final stages and for this we are grateful.

A special acknowledgment is due Sidney Johnson and J. W. (Pat) O'Meara of the former Office of Saline Water of the Department of the Interior. Both persisted throughout the fall of 1973 following the Arab oil embargo to obtain support for the work on which this book is based. At the beginning of 1974 they did succeed in having a small contract awarded to one of us (R.F.P.) to consider if future work on the problem was justified. The results of that study served as the basis for the program later undertaken by the National Science Foundation.

We are also grateful to a number of agencies and individuals from whom we have received indirect support through ongoing study contracts on the subject matter of this book. We wish in particular to thank the Industrial and Environmental Research Laboratory of the U.S. Environmental Protection Agency at Research Triangle Park, N.C., and T. Kelly Janes, James Chasse, and Dennis J. Cannon of that laboratory; the Office of Energy, Minerals and Industry of the U.S. Environmental Protection Agency and Gary Foley and Steven E. Plotkin of that office; the Division of Fossil Energy of the former U.S. Energy Research and Development Administration and Matthew J. Reilly, John H. Nardella, Hershul Jones, and Thomas Nakeley of that division. All of the individuals mentioned interacted strongly with our programs and provided much valuable advice and assistance.

A particular debt of gratitude is due David Goldstein of Water Purification Associates whose many contributions appear throughout the text. We are also grateful to all the other members of Water Purification Associates who provided us with technical assistance, comments, and reviews. R. Edwin Hicks, Li Liang, Joseph S. Shen, Irvine W. Wei, and David Yung all merit many thanks. We wish also to acknowledge Resource Analysis, Inc., and David H. Marks, Richard L. Laramie, and John H. Gerstle in particular, for supplying information on water resources that was developed under several joint programs with Water Purification Associates. This information was incorporated without reference in Sections 9.2 and 9.3. Stone and Webster Engineering Corporation assisted Water Purification Associates in some of their sponsored studies on the subject matter of this book, and we particularly acknowledge the contributions of Winthrop D. Comley and Carl H. Jones to the discussions on cooling and on water treatment.

For her preparation of the figures we are grateful to Nancy Holbrook. For the typing of the manuscript and the handling of the numerous secretarial details we thank Christine Stilton and P. Margaret Qamoos.

Finally, our most sincere thanks and appreciation go to Debra Knopman who through her copy editing greatly improved the style of the original manuscript.

Water in Synthetic Fuel Production

1
Introduction and Summary

1.1 Synthetic Fuels and Water

The recognition of declining world reserves of petroleum and natural gas has accelerated the development of the technology to produce synthetic fuels compatible with existing equipment. Synthetic fuels are obtained by converting a carbonaceous material to another form. In the United States the most abundant naturally occurring materials suitable for this purpose are coal and oil shale. The conversion of these raw materials may also be undertaken to remove sulfur or nitrogen that would otherwise be burned, giving rise to undesirable air pollutants. Another reason for conversion is to increase the heating value of the original coal or shale by removing unwanted constituents such as ash and thereby produce a fuel which is cheaper to transport and handle.

The manufacture of synthetic fuels from coal or oil shale may be regarded as a process of hydrogenation in which water is the source of the hydrogen. A typical bituminous coal by weight contains 75 percent carbon, 5 percent hydrogen, and 20 percent inert or undesirable matter. On the other hand, for example, synthetic natural gas which is almost entirely methane contains 75 percent carbon and 25 percent hydrogen. In addition to the water required as a source of hydrogen, water is also generally needed for other process steps.

As in any real process, conversion can never be completely efficient. The available energy or heating value of the coal or shale cannot be fully recovered in the synthetic fuel. This unrecovered thermal energy must be transferred to the environment in some way. Generally, part of the unrecovered heat is disposed of by evaporating water. Finally, water is needed to mine and prepare the raw material and to dispose of the unwanted constituents removed in the process.

In the United States, the importance of water usage in synthetic fuel production lies principally in the fact that much of the easily mined coal and almost all of the high grade oil shale are found in the arid western areas of the country. The nation's richest "hydrocarbon basin" cuts a swath from Montana and the Dakotas in the north down through Wyoming, Utah, and Colorado to New Mexico in the south. The center of the coal and shale swath underlies the Upper Colorado River Basin, with the remainder encompassing part of the Missouri River Basin in the north and the Rio Grande Basin in the south. In these arid, sometimes drought-ridden areas, agricultural, industrial, municipal, recreational, and power needs compete for a limited water supply. Even now a *de facto* overcommitment of available fresh water resources exists in some portions of the hydrocarbon basin. The problem of water shortage is, however, not limited to the

West. In the humid coal areas of Illinois and Appalachia in the East, local and temporal water shortages may be a major impediment to the development of synthetic fuel facilities at coal mining sites. In all regions, the water discharge problem is constrained by the environmental limitation that the plant effluent waters may not add pollutants to the surface or ground water or otherwise disrupt the aquatic equilibrium.

Published estimates vary widely on the amount of water consumed in manufacturing synthetic fuels. Generally, the reason for this variance is the amount of water that is assumed to be evaporated for cooling in the plant, since cooling water is most often the prime determinant of total consumption. In conversion processes the heat to be removed will be less, the higher the efficiency of the conversion. For this reason, water usage in coal liquefaction will tend to be somewhat less than in coal gasification, when measured in relation to the thermal output of the product fuel.

By itself, the criterion of conversion efficiency may not suffice to characterize the water consumption. For example, one gasification process may consume considerably less water than another because the moisture in the coal is collected and used, although the utilization of such dirty water will not be without cost. To provide a reference point, consider a mine-plant complex to gasify coal. A plant that is reasonably well designed so as not to be wasteful of water, but not designed to minimize water consumption, might require about 18 gallons of water per million Btu of heating value in the gas. For a standard size plant producing 250 million cubic feet of pipeline gas per day, this amounts to about 4.5 million gallons of water per day. This value may be compared with a commonly quoted range of 9 to 40 million gallons per day.[1]

If synthetic fuel development is to be a viable prospect in regions where water is in short or uncertain supply, then effluent process waters must be reused, water consumed for cooling must be minimized, and unusual water sources such as brackish groundwater and municipal effluents may have to be used. The actual amount of water consumed depends on a number of factors including the product fuel, the cooling method, the site, the process, and the methods of disposing of the residuals. However, there is no absolute water requirement but only a preference since it is at the discretion of the designer to reduce the quantity to very low levels. The preference is principally an economic one that largely depends on the site as it affects the real cost of water. Of course, the preference may also rest on social, political, or environmental grounds. In this book the technology and alternatives to minimize water consumption and pollution are described.

1.2 Summary and Conclusions

The synthetic fuel technologies examined include the conversion of coal to clean gaseous, liquid, and solid fuels, and the conversion of oil shale to clean liquid fuels. A number of processes are described for each conversion, including both above ground and *in situ* (underground) procedures. For purposes of comparing water requirements, water treatment plants, and residuals generated, detailed conceptual designs for integrated mine-plant complexes are presented for representative conversion processes. Only above ground processing is considered in detail. The processes and products chosen for purposes of comparison are shown in Table 1.1. Specific designs are based on standard size plants with the given product outputs.

The coal mining regions chosen were those where the largest and most easily and economically mined deposits are located. In the West these include the Powder River and Fort Union regions in Montana, Wyoming, and the Dakotas, and the Four Corners region where New Mexico, Arizona, Utah, and Colorado meet. In the central and eastern areas of the country, the Illinois and Appalachian basins were selected. The western coals are principally low sulfur subbituminous and lignite, while the eastern coals are mainly high sulfur bituminous. Average heating values for the different coal ranks are: bituminous 13,000 Btu/lb; subbituminous 9,800 Btu/lb; lignite 6,800 Btu/lb. Only high grade oil shale from the Green River Formation is considered. Specific design examples are restricted to shales with yields of about 30 to 35 gallons per ton, as might be found in Colorado or Utah.

The plant output measured in terms of the heating value of the product fuel

Table 1.1
Conversion processes, products, and plant sizes
compared for water requirements.

Technology	Conversion Process	Product	Standard Size Plant Output
Coal Gasification	Synthane, Hygas, Lurgi	Pipeline Gas	250×10^6 scf/day*
Coal Liquefaction	Synthoil	Fuel Oil	50,000 barrels/day
Clean Coal	SRC	Solvent Refined Coal	10,000 tons/day
Oil Shale	Paraho Direct and Indirect, TOSCO II	Synthetic Crude	50,000 barrels/day

*Standard cubic feet per day.

Table 1.2
Process efficiencies and mining rates for standard size synthetic fuel plants.

Technology	Heating Value of Product Fuel (10^{11} Btu/day)	Process Conversion Efficiency (percent)	Coal or Shale Mining Rates* (10^3 tons/day)
Coal Gasification	2.4	65–70	16–30
Coal Liquefaction	3.1	70–75	16–31
Clean Coal	3.2	75–80	17–33
Oil Shale	2.9	57–72	75–100

*Highest coal mining rates for lignite, lowest rates for bituminous. All shale high grade (30 to 35 gal/ton); highest shale mining rate for Paraho Indirect process, lowest rate for TOSCO II process.

together with the process conversion efficiency and heating value of the raw fuel roughly define the coal and shale mining rate. Table 1.2 gives an approximate range of overall process conversion efficiencies, defined by the sum of the heating values in the product fuel and byproducts compared to the heating value in the raw coal or shale. For the coal conversion plants, Table 1.2 summarizes the plant sizes and mining rates for the range of heating values of the different rank coals. The range in shale mining rates corresponds to the different conversion efficiencies of the processes considered.

For a given size coal conversion plant, the quantity of water consumed depends mainly on four factors: the product, the fraction of unrecovered heat disposed by wet cooling, and to a lesser extent the site and specific process. In the above ground conversion of oil shale to synthetic crude there are three important factors affecting water consumption: the method of spent shale disposal, the shale retorting process, and the extent to which wet cooling is used.

Estimates of water consumption are net ones; all effluent streams are assumed to be recycled or reused within the mine or plant after any necessary treatment. These streams include the dirty waters generated in the conversion process and the highly saline water drawn off (blown down) from wet evaporative cooling towers. Water only leaves the plant as vapor, as hydrogen in the hydrocarbon products, or as occluded water in the solid residues. Dirty water is cleaned, but only for reuse and not for returning it to a receiving water.

The process water requirements relate to the fact that hydrogen is needed for the fuel conversion, and its source is water. The water consumed depends on the difference between the hydrogen to carbon ratio in the coal or shale and this same ratio in the product fuel plus byproducts. A second factor in determining net consumed process water is the moisture present in the coal or shale, which is not treated as a water input but which may be recovered in the process. The net process water requirement is generally not large and water may even be produced; it represents the difference between the high pressure steam needed for the process and the dirty water generated in the process. However, each of these streams by themselves may be quite large. Moreover, the qualities of these streams are at opposite ends of the spectrum: to produce the steam it is necessary to treat the boiler feed water to very high purity, while the water generated by the process is highly contaminated with organic matter, ammonia, and hydrogen sulfide.

The quantity of water consumed in cooling depends principally upon the overall plant conversion efficiency. This specifies the heat that is not recovered and that must be dissipated to the environment. Not all of this unrecovered heat must be dissipated by cooling, as an appreciable fraction of it will be lost directly to the atmosphere. And disposing of all the remaining unrecovered heat by evaporating water may not be economical; some of the heat should be transferred to the atmosphere by forced air cooling. The extent to which water is evaporated to remove the rest of the heat depends upon whether the region in which the plant is located is water rich or water short, coupled with the true cost of water. Approximately 1,400 Btu's are transferred for every pound of water evaporated in a wet evaporative cooling tower. In water rich areas, typically 25 to 60 percent of the total unrecovered heat goes to evaporating water, while in arid regions or where water is expensive only 10 to 30 percent of the unrecovered heat is used to evaporate water.

The water required for mining and preparation of the coal or shale and for the disposal of ash or spent shale is a function of location, principally through the amount of material that must be mined or disposed. Sulfur removal also consumes water, and the amount depends not only on the coal but also on the conversion process. Water is also needed for a number of other purposes that depend upon climate, such as land reclamation. Generally, any one of these requirements is not large and the needs can be met with lower quality waters. Nevertheless, when the requirements are taken together they are significant and cannot be neglected in any plant water balance, although general rules for the amount consumed are not easily stated. Differences in consumption in this category for a given coal conversion process, however, do not vary by more than 15 percent between regions with the exception of the Four Corners region. The difference is somewhat greater when this region is compared since larger amounts of water are needed there for ash handling, dust control, and revegeta-

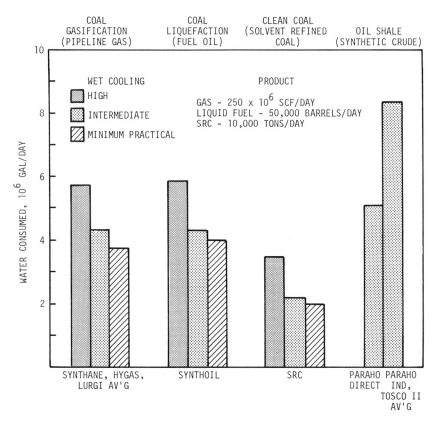

Figure 1-1
Net water consumption for standard size synthetic fuel plants.

tion. Figure 1-1 shows the total water consumed in standard size synthetic fuel plants with the regional differences averaged out for each coal conversion technology.

The differences in consumptions indicated in Fig. 1-1 for each of the coal conversion processes are a function of the degree of wet cooling assumed to be used in disposing of the unrecovered heat. The term high wet cooling means that forced air coolers are used only where clearly cheaper than evaporative wet cooling even when water has no value. This occurs when cooling high temperature streams, usually above 140°F. Intermediate and minimum practical wet cooling correspond to additional use of combined wet evaporative cooling towers and forced dry air coolers. This can reduce the cooling water consumption by one half to two thirds. The rather large savings in water achievable through the use of a combined wet and dry cooling system can be carried out at a cost not likely to exceed 1 percent of the sale price of the product fuel. For the high wet cooling system, by far the largest single consumptive use of water in any plant is for cooling, ranging between 2 million and 4 million gallons per day for all of the processes.

In oil shale conversion only one degree of wet cooling is indicated. Here in the TOSCO II process the largest consumptive use of water is for spent shale disposal, amounting to 50 percent of the total consumption. For the Paraho Direct process the corresponding fraction is 25 percent, with cooling water accounting for 60 percent of the total consumption.

The absolute water consumptions indicated in Fig. 1-1 for the standard size

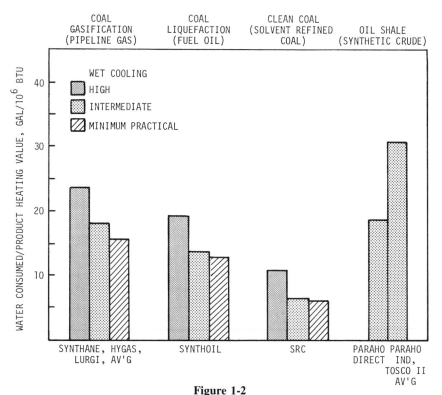

Figure 1-2

Water consumption in synthetic fuel production normalized with respect to the product fuel heating value.

plants show a variation between 2 million and 8 million gallons per day, depending on the process and degree of wet cooling. The lowest consumptions are for solvent refined coal production, while the highest are for producing synthetic crude from oil shale. In Fig. 1-2 these water requirements are normalized with respect to the heating value in the product fuel, thus enabling a more direct comparison between processes. Among the coal conversion processes, irrespective of the wet cooling option, on a normalized basis, solvent refined coal production consumes the least water and pipeline gas production the most water. These results show that the higher is the overall conversion efficiency the lower is the water consumption.

For the averages of the pipeline gas processes, the difference in consumed water between the high wet cooling and minimum practical wet cooling options is about 8 gal/10^6 Btu or 2 million gallons per day. For individual processes the difference is even larger; it is about 12 gal/10^6 Btu or about 3 million gallons per day for the Synthane process. The difference in normalized water requirements by a factor of almost four between the high wet cooling case for pipeline gas production and the minimum cooling case for solvent refined coal production points up the importance of the choice of product and cooling design in the amount of water consumed in synthetic fuel production.

To convert oil shale to synthetic crude by the Paraho Direct process requires about the same amount of water as does the conversion of coal to pipeline gas. The larger water requirement for the TOSCO II process relates to the great amount of water needed to dispose of the spent shale. For the Paraho Indirect process the larger water requirement is a consequence of the relatively low overall conversion efficiency.

In all estimates of water consumption, reuse or recycling of water is assumed, with the required water treatment plant integrated with the synthetic fuel complex. For the highest level of treatment required to minimize the consumption of water, the complete water treatment cost including amortized capital costs should be in the range of $0.02 to $0.10/$10^6$ Btu in the product fuel. This cost is not likely to exceed 5 percent of the sale price of the product fuel from any of the plants.

Every synthetic fuel plant generates solid residuals, most of which are disposed wet with occluded water that is not economical to recover. In the coal conversion plants these residuals are principally the coal ash and, where sulfur dioxide must be removed from stack gases, the flue gas desulfurization sludge. In oil shale plants the principal residual is the spent shale. Figure 1-3 shows the amount of residuals generated by the standard size plants and Fig. 1-4 these same

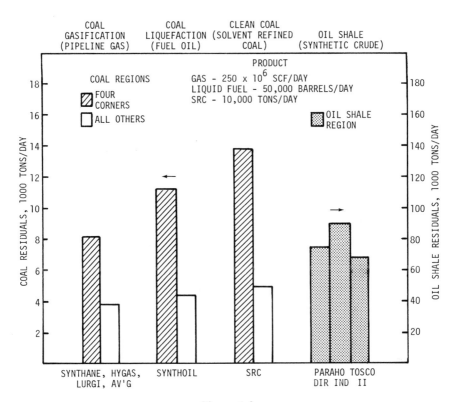

Figure 1-3
Wet solid residuals generated in standard size synthetic fuel plants.

quantities normalized with respect to the heating value in the product fuel. For each coal conversion technology, the residual quantities have been averaged for all the coal regions except Four Corners; the differences among these regions are not large. In Four Corners, on the other hand, the ash content of the coal is more than 25 percent for the coal considered, and for this reason the residual quantities are shown separately. It is of interest that for all the technologies and most of the coals, the total residuals generated are in a quite narrow range around 30 lbs/10^6 Btu in the product fuel. Outstripping all of the coal conversion residuals by an order of magnitude is the spent shale from oil shale processing.

Estimates of water availability in the principal coal and oil shale regions are made on the basis of supply and demand requirements including water used or allocated for agricultural, industrial, and municipal uses as well as other forms of energy development. A representative level of synthetic fuel development may

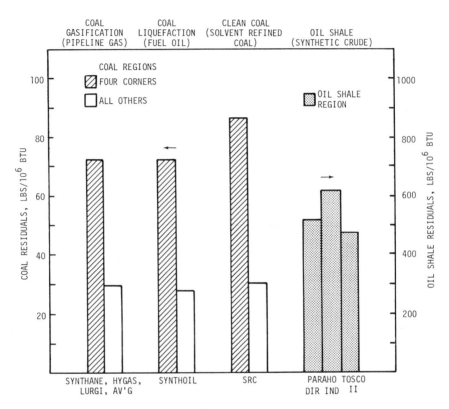

Figure 1-4
Wet solid residuals generated in synthetic fuel production normalized with respect to the
product fuel heating value.

be considered an industry producing 1×10^6 barrels/day of synthetic crude or its
equivalent of 5.8×10^{12} Btu/day in other fuels, in each of the five principal coal
bearing regions and in the principal oil shale region. This corresponds to a total
production of 6×10^6 barrels/day of synthetic crude or its equivalent in other
fuels. For the standard size plants examined, 1×10^6 barrels/day of synthetic
crude or its equivalent in heating value is produced from 20 oil shale plants, 19
coal liquefaction plants producing fuel oil, 18 solvent refined coal plants, or 24
pipeline gas plants. Figure 1-5 shows the percentage of the average annual values
of available surface water that would be consumed in each region by this level of
synthetic fuel development. The Appalachian and Illinois basins are combined in
this figure and represent a production level of 2×10^6 barrels/day. Because of the

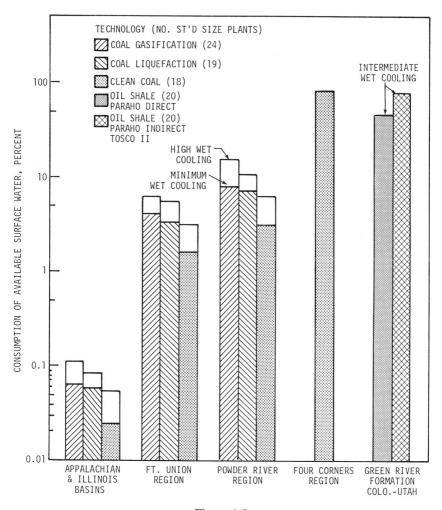

Figure 1-5
Water consumption as a percentage of the average annual values of available surface water
to produce 1×10^6 barrels/day of synthetic crude or its equivalent of 5.8×10^{12} Btu/day
in each of six regions, for a total production of 6×10^6 barrels/day.

limited water supply in the Four Corners region, only solvent refined coal production could be met by the available surface water supply. For the production
level assumed, the other technologies exceed the average annual values of available surface water in this region and therefore are not shown.

From the criterion of water availability alone, Fig. 1-5 indicates that except for the most arid areas and those areas where water is already largely allocated, a relatively high level of synthetic fuel production could be supported in the principal coal and shale regions of the United States.

Chapter 1. References

1. Tetra Tech, Inc., "Energy From Coal—A State-of-the-Art Review," ERDA Report No. 76-7, U.S. Gov't. Printing Office, Washington, D.C., 1976.

2
General Considerations

2.1 Hydrogenation and Other Process Water

The weight ratio of carbon to hydrogen (C/H) is higher in coal and shale than in gaseous or liquid synthetic fuels, so that the conversion can be regarded as a process of hydrogenation. If water is the source of hydrogen then the minimum amount of water necessary to convert the coal or shale to a synthetic fuel may be determined. This minimum water corresponds to the increase in hydrogen content of the product fuel over that in the raw material. Table 2.1 shows the minimum theoretical water quantity required to convert coal to a variety of fuels in relation to their carbon to hydrogen weight ratio.

In any coal or shale conversion the method of hydrogenation may be direct, indirect, or by pyrolysis, either alone or in combination. Direct hydrogenation involves exposing the raw material to hydrogen at high pressure. Usually, the hydrogen is produced by reacting steam with carbon char. Indirect hydrogenation involves reacting the raw material directly with steam or dissolving the raw material in a hydrogen donor solvent. In pyrolysis the coal or shale is heated in an inert atmosphere and then decomposes to yield solid carbon, and gases and liquids with higher fractions of hydrogen than the original material. Pyrolysis is generally a first step in most conversion processes. Any water vapor that distills off during pyrolysis is condensed and the oil fractions separated out. This water, which we term process condensate, is quite dirty and must be treated either for reuse or discharge.

To obtain fuels that will burn cleanly, sulfur and nitrogen compounds must be removed from the gaseous, liquid, and solid products. Sulfur mostly in the form of hydrogen sulfide and nitrogen in the form of ammonia are present in the gas made from coal or the pyrolysis of shale and also in the gas generated in the hydrotreating of pyrolysis oils and synthetic crudes. To remove the hydrogen sulfide and ammonia may require the use of water as a liquid wash, while steam may be needed for heating or for gas stripping. Sulfur is also present as sulfur dioxide in the combustion gases when coal or char is burnt to generate process steam. The removal of this sulfur dioxide generally involves a considerable amount of water.

Although there are other process water uses, those for hydrogenation and the removal of contaminants from the synthetic fuels are the principal ones. The principal dirty process waters are what we have termed the process condensates and those effluents from fuel cleaning and purification. In our process water definitions we exclude cooling water requirements and the resulting dirty blowdown waters, choosing to treat these needs and effluents separately.

Table 2.1
Minimum water requirements to convert coal to typical fuels.

Fuel	C/H Weight Ratio	Molar Representation	Minimum Conversion Water (gal/10^3 lb product)
Coal	~ 15	$CH_{0.8}$	—
Benzene	12	$CH_{1.0}$	17
Crude Oil	~ 9	$CH_{1.33}$	43
Gasoline	6	CH_2	93
Methane	3	CH_4	216

2.2 Cooling

Cooling, or the dissipation of heat, is an essential part of any synthetic fuel process. Heating value in the raw material not recovered in the synthetic fuel or the byproducts must be transferred to the environment. Some of this unrecovered heat is lost directly to the atmosphere, leaving the plant in hot gases up a flue, in water vapor from coal drying, in convective and radiant losses from machinery and container surfaces, and in other direct ways. However, the largest fraction of the unrecovered heat is indirectly transferred to the atmosphere, or sometimes to a large body of water, through a heat transfer surface. To the designer concerned with water usage, this indirect heat transfer is most important, since it is here where water can be consumed in large quantities.

The heat to be dissipated may be transferred through a heat transfer surface that is cooled by air. This is called dry cooling or air cooling, and negligible water consumption is incurred in this method. Alternatively, the heat transfer surface may be cooled by circulating water, which is itself cooled in a cooling tower with evaporation to the atmosphere of a fraction of the water. This is called wet cooling or evaporative cooling. We shall be concerned in the book primarily with these two methods of cooling or their use in combination, termed wet/dry cooling.

Under the same conditions, an air cooled heat exchanger transfers less heat per unit area than a water cooled exchanger. Therefore, to transfer a given amount of heat, air cooling requires a larger surface area. Air coolers thus have a higher capital cost. Moreover, an air cooled system cannot usually reach temperatures as low as those in a water cooled system, and hence the efficiency of conversion will be lower if air cooling is used to any great extent.

Wet cooling, however, does require water, and water is not free. To move one thousand gallons of water one mile through a horizontal pipeline costs 1 to 2 cents. To treat water in a circulating cooling system and to dispose of the residues can cost $0.50 to $1.00 per thousand gallons of water evaporated.[1] The economic decision of whether to use wet or dry cooling or a combination at any given point of cooling in the plant depends on the true cost of water. This must include any costs for buying water rights and transporting the water to the plants, as well as the cost of treating the circulating cooling water and disposing of the cooling tower residues.

Two other common methods of cooling that should be mentioned are once-through cooling, in which the cooling water is discharged directly back to the source from which it is taken, and pond cooling in which the cooling water is discharged into large ponds from which the heat is transferred to the atmosphere by radiation, convection, and evaporation. Once-through cooling systems presuppose that large quantities of water are available, which in the West and today even in the East is rarely the case. In any event, the environmental consequences of discharging heated water will markedly limit this form of cooling in the future. For these reasons, we shall not consider this method in any detail. Cooling ponds require relatively large land areas and generally consume more water than wet cooling. They are, however, inexpensive and relatively easy to maintain. Cooling ponds will be discussed in Chapter 4 when the topic of cooling is examined.

2.3 Mining, Fuel Preparation, and Residuals Disposal

In any mine-plant complex for synthetic fuel production, water will be consumed in mining the coal and shale and reclaiming the mined land, in preparing the raw materials for the conversion process, and in disposing of the residuals. The water needs associated with the mining of the coal or shale will not be strongly affected by the conversion process, provided the conversion takes place above ground, except as it determines the actual quantity of material to be mined. However, the geographical location of the mine and whether the mining is surface or underground will influence the quantity of water consumed.

A principal mine water need is for dust control at the mine, in the surrounding areas, and on the mine roads. Water is also required in holding down dust in the handling and crushing of the coal or shale. Coal washing for the purpose of removing ash or sulfur also consumes water. We shall, however, assume that coal washing generally will not be used in synthetic fuel mine-plant complexes

for reasons to be discussed in Chapter 7. The requirement of water for supplemental irrigation to reclaim strip-mined land can be large but should be necessary
only in the most arid regions.[2]

All residuals, such as ash in the coal or salts in the water, that enter the plant
boundaries must be disposed of along with residues from materials added in the
processing steps. This may require significant quantities of water. Ash can represent as much as one-third the weight of the incoming coal. For most coals, ash
ranges between 5 and 15 percent by weight, although in some western coals it is
as high as 25 to 30 percent. The most common method of disposing of this ash
involves mixing it with water and slurrying it to a disposal site. With shale the
problem of disposal is even more severe, for about 80 to 85 percent by weight of
the originally mined oil shale remains as spent material, with a volume before
compaction averaging 50 percent greater than its in-place volume, and even after
maximum compaction at least 12 percent greater.[3] Large volumes of water may
be needed for the shale compaction process and are needed for the subsequent
revegetation of the spent shale piles.

Other uses of water involve the non-specific mine and process needs including
service, sanitary, and potable water consumption. Similar requirements will be
associated with any satellite town allied with the mine-plant complex. Finally,
since all the water that is used will generally have come from an onsite reservoir
and since it may have passed through settling basins, it is also necessary to
determine the evaporation losses from these holding areas. These losses are
chargeable to the water requirements for the mine-plant complex.

2.4 Coal and Shale Deposits

If synthetic fuel plants are to be sited where the coal and shale is mined, then the
location of these deposits is important in determining not only the water
availability but also the water consumption. The coal fields of the conterminous
United States and the rank of coal found in these fields are shown in Fig. 2-1.
Coal rank refers to the percentage of carbon and heat content of the coal. The
fraction of carbon in the coal increases from lignite to anthracite, and the
moisture fraction decreases. The fact that the coal moisture varies considerably
with the type of coal can affect the process water requirements in a synthetic fuel
plant. For a given synthetic fuel output, the heating value of the coal determines
the actual quantity of coal required and in this way influences the water needs.
The heating value increases from lignite to low-volatile bituminous coal. Anthracite has a heating value about equal to that for medium-volatile bituminous coal.

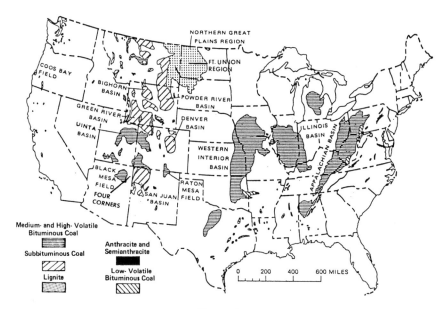

Figure 2-1
Coal fields of the conterminous United States.

The demonstrated coal reserve base distributed according to region is shown in Table 2.2, compiled from the data of Averitt.[4] This reserve refers to identified resources suitable for mining by present methods, where at least 50 percent is recoverable. The coal in this category lies less than 1000 feet below the surface. Table 2.2 also shows the potential methods by which the coal can be mined. In the Northern Great Plains and Rocky Mountain region, where almost half of the

Table 2.2
Demonstrated coal reserve base of the United States in billions of tons
by region and potential method of mining.

Region	Underground	Surface	Total	Percent of Grand Total
Northern Great Plains and Rocky Mountain	113	86	199	46
Appalachian Basin	97	16	113	26
Illinois Basin	71	18	89	20
Other	16	17	33	8
Grand Total	297	137	434	100

nation's coal is to be found, more than 40 percent of the coal can be surface mined. Surface or strip mining can be done more economically and in most cases with a much higher proportion of the coal recovered.

Oil shale is a rock, generally of a sedimentary type, containing organic matter which when heated to its pyrolysis temperature yields a certain minimum quantity of oil. The organic matter contains an insoluble component termed "kerogen," which yields most of the oil. When the shale is heated to its pyrolysis temperature of around 900°F, oil, gas, and water vapor, the latter in amounts

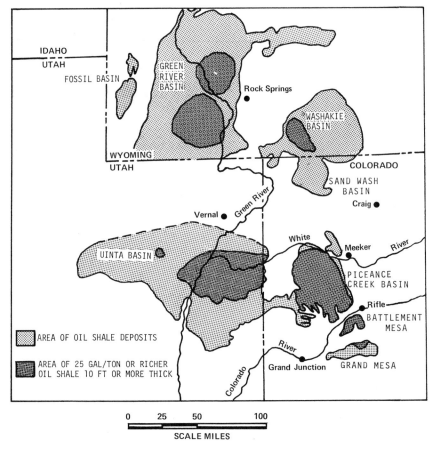

Figure 2-2
Oil shale areas of the Green River Formation in Colorado, Utah, and Wyoming.

comparable to that of the oil, will distill off, leaving behind a spent residue mainly composed of inorganic matter. Shale with about 7 percent by weight of organic matter yields 10 gallons of oil per ton of shale. This is considered to be the lower limit for the shale to be classified as oil shale. High grade oil shale is taken to be shale with an organic content greater than about 14 percent that yields 25 gallons or more of oil per ton of shale and is found in beds at least 10 feet thick. It should be noted that oil shale deposits tend to be much thicker than coal seams and that the shale is considerably harder than coal.

Large amounts of lower grade shale are found in many areas of the United States, but particularly in the same regions as the coal basins of the East and Midwest (see Fig. 2-1). However, the greatest promise for commercial production lies where the high grade oil shale is found, in areas in Colorado, Utah, and Wyoming underlain by what is called the Green River Formation. Figure 2-2 outlines the oil shale areas in this formation.[3] The identified high grade shales with yields between 25 and 65 gallons per ton have an oil equivalence of about 570 to 620 billion barrels. The richest, most accessible zones are estimated to yield about half this amount.[5] About 80 percent of the high grade material is located in Colorado in the Piceance Creek Basin.[5,6] These deposits are in strata varying from 10 to 2000 feet in thickness, at depths to 1600 feet below the surface and in outcroppings at the surface.[3]

The identified lower grade shale yield is more than three times that for the higher grade shale. Only the processing of the high grade oil shale will be emphasized in this book, since the economics of converting the lower grade material is considerably less promising.

The mining of oil shale for surface processing is roughly the same as for coal. Underground and surface mining is envisaged, although the amount of shale that can be economically open-pit mined is estimated to be only about 10 to 15 percent of the total.[7]

2.5 Water Availability and Constraints

The availability of water is very specific to location. Although regional generalizations are possible, no analysis of water supply for a synthetic fuel mine-plant complex can be made without a detailed knowledge of the local supply and demand characteristics. Among the important features of the local supply are the seasonal and annual reliability, the transport and storage requirements, the source (i.e., surface or groundwater), and the seasonal variations in water quality. For example, it is likely that it generally will not be economic to pipe water to a mine-plant complex a distance much over 100 miles from the source of supply. Among the important competing demands for the supply are those from irrigated agriculture, from municipal and recreational uses associated with increased population in the coal and shale mining areas, from steam-electric power generation and other energy uses, and from industrial needs associated with economic growth.

Environmental and institutional constraints affecting water supply must also be taken into account. Principal among environmental considerations are the possibility of the disruption of natural underground reservoirs from the mining operation and surface and groundwater contamination from the leaching of disposed wastes or from acid mine drainage. The institutional constraints include the legal doctrines governing the use of water. In the East this is generally the Riparian Doctrine, which defines surface water rights as ownership of land next to or traversing the natural stream. In the West the Appropriation Doctrine usually applies: the first appropriation of the water conveys priority, independently of the location of the land with respect to the water. Other constraints may involve competing claims, such as Indian water rights.

With the issues noted above in mind, we can roughly characterize the principal water supply considerations in the major coal and shale regions. In the Appalachian and Illinois coal basins where mine-plant complexes could be located, a sufficient and reliable supply of surface water exists close to most of the major streams. In the Appalachian Basin, however, the surface supplies are much less reliable in the upper water courses encompassing part of the eastern Kentucky and West Virginia coal ranges. In many instances riparian land requirements can prevent the transfer of surface water even over a short distance to a non-riparian mine-plant site, thereby causing an otherwise adequate supply to be inaccessible. Groundwater supplies appear to be available where surface supplies may be inadequate or questionable. Acid mine drainage still presents problems throughout the Appalachian Basin.

In the Fort Union and Powder River coal regions there appears to be an adequate surface water supply, which includes the large flows of the Upper Missouri and Yellowstone Rivers and, in addition, a considerable storage capacity. However, chronic local water shortages do exist, especially in the northern Wyoming area of the Powder River Basin. Moreover, during periods of low flow, water shortages occur in parts of the Yellowstone Basin. Water is obtainable by appropriation and can be transferred by trans-basin diversions. There are, however, a number of serious institutional conflicts in the region, particularly in Montana and Wyoming, concerning the authority to allocate the water. Competitive pressures from agricultural water users are very high and irrigation needs large because of the semi-arid character of the area. Environmental problems associated with the disruption of natural underground reservoirs by mining may also be important.

The Four Corners coal region and Green River Formation shale area encompassing the Colorado and Rio Grande River Basins is an arid area marked by an inadequate water supply of poor quality. The region is subjected to highly variable annual stream flows and multi-year periods of high and low flows. Water is a limiting factor in the development of the region and sufficient quantities are not available to meet the needs for projected synthetic fuel programs. It may be possible to utilize groundwater as a conjunctive supply, but this water is generally of poor quality and is often drawn from unreplenished underground reservoirs, which would eventually be depleted. Strong competition exists among agricultural, municipal, and industrial users for the available supply, most of which is now either appropriated or overappropriated. Serious institutional conflicts involving Indian water rights also exist in the area. For any large scale synthetic fuel development it appears that major storage projects would be needed to provide a dependable water supply.

No mention has been made of the consequences of periods of drought which periodically afflict all regions of the country, but any evaluation of water availability for synthetic fuel development must also take this possibility into consideration.

Chapter 2. References

1. Gold, H., Goldstein, D. J., and Yung, D., "Effect of Water Treatment on the Comparative Costs of Evaporative and Dry Cooled Power Plants," Report No. COO-2580-1, Div. of Nuclear Res. & Application, Energy Res. & Develop. Admin., Washington, D.C., July 1976.

2. National Academy of Sciences, *Rehabilitation Potential of Western Coal Lands*. Ballinger Publishing, Cambridge, Mass., 1974, p. 168.

3. U.S. Department of the Interior, "Final Environmental Statement for the Prototype Oil Shale Leasing Program," Vol. I, U.S. Gov't. Printing Office, Washington, D.C., 1973.

4. Averitt, P., "Coal Resources of the United States, January 1, 1974," Geological Survey Bulletin No. 1412, U.S. Gov't. Printing Office, Washington, D.C., 1975.

5. Hendrickson, T. A., *Synthetic Fuels Data Handbook*. Cameron Engineers, Inc., Denver, Colo., 1975.

6. Keighin, D. W., "Resource Appraisal of Oil Shale in the Green River Formation, Piceance Creek Basin, Colorado," *Quarterly Colorado School of Mines* **70** (3), 57–68 (1975).

7. Schmidt-Collerus, J. J., "The Disposal and Environmental Effects of Carbonaceous Solid Wastes from Commercial Oil Shale Operations," Report No. NSF-RA-E-74-004 (NTIS Catalog No. PB 231 796), Denver Research Institute, Denver, Colo., Jan. 1974.

3
Coal and Shale Conversion Fundamentals

3.1 Introduction

The process of converting coal or shale to a synthetic fuel has been described as basically one of hydrogenation, in which the weight ratio of carbon to hydrogen is higher for the raw material than for the liquid or gaseous synthetic fuel. In the conversion, sulfur and nitrogen are reduced to produce a cleaner fuel, and ash, oxygen, and nitrogen are reduced to produce one with a higher heating value.

Synthetic fuels include low-, medium-, and high-Btu gas; liquid fuels such as fuel oil, diesel oil, gasoline; and clean solid fuels. Low-Btu gas, often called producer or power gas, has a heating value of about 100 to 250 Btu per standard cubic foot (written as Btu/scf). This gas is an ideal turbine fuel whose greatest utility will probably be in a gas-steam combined power cycle for the generation of electricity at the location where it is produced. Medium-Btu gas is loosely defined as having a heating value of about 250 to about 550 Btu/scf, although the upper limit is somewhat arbitrary, with existing gasifiers yielding somewhat lower values. This gas is also termed power gas or sometimes industrial gas. It may be used as a source of hydrogen for the production of methanol and other liquid fuels. It may also be used as a fuel for the production of high-Btu gas, which has a heating value in the range of about 920 to 1000 Btu/scf, and is normally composed of more than 90 percent methane. Because of its high heating value, high-Btu gas is a substitute for natural gas and is suitable for economic pipeline transport. For these reasons it is often referred to as substitute natural gas or pipeline gas, as well as synthetic natural gas.

The synthetic gases can be produced from coal by indirect hydrogenation in which the gasification takes place by reacting steam with the coal, or by direct hydrogenation, termed hydrogasification, in which hydrogen is contacted with the coal (see Fig. 3-1, adapted from Refs. 1 and 2). Oil shale is generally not thought of as a primary raw material for gas production, although it is possible to use and has been discussed.[3]

Clean synthetic liquid fuels can be produced by several routes, as shown in Fig. 3-1. For example, coal can be gasified first and then the liquid fuel synthesized from the gas. This is not a very efficient procedure, since it involves breaking the carbon bonds in the coal and then putting some of them back together again. Another procedure is pyrolysis, the distillation of the natural oil out of the coal or shale. The oil vapors are condensed, the resulting "pyrolysis oil" is treated with hydrogen, and the sulfur and nitrogen in it is reduced. This is similar to the procedure used in "upgrading" crude oil in a refinery to produce a variety of liquid fuels. In the last procedure, the coal is dissolved in a hydrogen

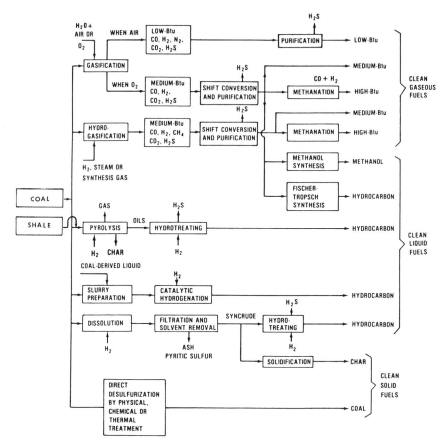

Figure 3-1
Methods of producing clean synthetic gaseous, liquid, and solid fuels.

donor solvent, and the ash and the physically separable pyritic sulfur are filtered out. After the solvent is removed, the resulting heavy synthetic crude oil is cleaned and hydrotreated by refinery procedures to upgrade it to the desired liquid fuels. If instead of hydrotreating, the synthetic crude is cooled down, then a solid, relatively clean fuel termed "solvent refined" coal is obtained. Of course, any physical or chemical procedure to clean the coal also results in a cleaner solid fuel.

3.2 Properties of Coal and Shale

Coal

In the synthetic fuel conversion processes, a number of properties of the raw fuel are important for specifying the most appropriate conversion method, characterizing the products and determining the clean water consumption and dirty water production.

Coals are broadly classified by rank, a measure of their fixed carbon content and mineral- or ash-free heat content. The coal of lowest rank is lignite, followed in increasing rank by subbituminous coal, bituminous coal, and anthracite. Generally, the lower the coal rank, the lower its fixed carbon content and heat content, and the higher the fraction of moisture and volatile matter. The heat content and the proximate analysis defining the fraction of moisture, volatile matter, ash, and fixed carbon are important properties of coal which must be known to characterize the conversion process. In Fig. 3-2 these properties are shown for coals of different rank.[4] The proximate analyses are presented on an ash-free basis. The ash content of coals in the United States ranges from about 2.5 to 33 weight percent and averages about 9 percent.[4]

To obtain a proximate analysis, coal is first dried at about 220°F. The loss in weight is recorded as moisture. The coal is then heated to a higher temperature of around 950°F and the further loss in weight is recorded as volatile matter. Finally, the coal is oxidized (burned) and the loss in weight is the fixed carbon. The residue is ash, the inorganic mineral matter.

More detailed properties of the coal are given by an ultimate analysis, a breakdown according to the most important elements. Typical ultimate analyses are shown in Table 3.1 for coals from the major fields outlined in Fig. 2-1. There are several ways of presenting an ultimate analysis. In Table 3.1 the analyses are given on an "as-received" (not dried) basis. This is the form most convenient for this book. Also, the moisture has been shown separately, which is different from some presentations where the moisture is shown by adding its hydrogen and oxygen contents to the hydrogen and oxygen in the dry coal.

The heating value of the coal can be estimated from the ultimate analysis using the Dulong Formula:[5]

$$Q_{HHV} \text{ (Btu/lb)} = 145.4C + 620 \left(H - \frac{O}{8}\right), \qquad (3.1)$$

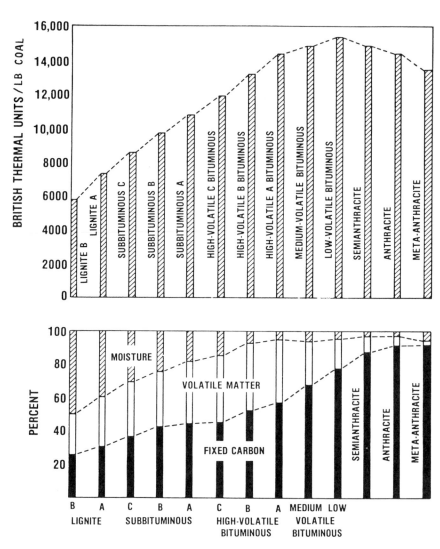

Figure 3-2
Heat content and proximate analyses of ash-free coals of different rank.

where C, H, and O are, respectively, the weight percentages of carbon, hydrogen, and oxygen in the dry coal. Here, the small contribution of the heating value of the sulfur has been neglected. The heating value Q_{HHV} is the gross or higher heating value (HHV) and includes heat released by condensation of steam formed

Table 3.1
Ultimate analyses in weight percent of representative coals of the United States.

Component	Fort Union Lignite	Powder River Subbituminous	Four Corners Subbituminous	Illinois C Bituminous	Appalachia Bituminous
Moisture	36.2	30.4	12.4	16.1	2.3
Carbon	39.9	45.8	47.5	60.1	73.6
Hydrogen	2.8	3.4	3.6	4.1	4.9
Nitrogen	0.6	0.6	0.9	1.1	1.4
Sulfur	0.9	0.7	0.7	2.9	2.8
Oxygen	11.0	11.3	9.3	8.3	5.3
Ash	8.6	7.8	25.6	7.4	9.7
Total	100	100	100	100	100
Higher Heating Value (Btu/lb)	6,720	7,920	8,440	10,700	13,400

during combustion. However, this heat of condensation is normally not recovered in combustion, and thus the net or lower heating value (LHV) is applicable. The lower heating value may be estimated from the relation

$$Q_{LHV} = Q_{HHV} - 92.7 \, H. \tag{3.2}$$

The higher heating values shown in Table 3.1 are close to those given by Eq. (3.1). They also agree with the heating values in Fig. 3-2 when those values are appropriately reduced to account for the ash content.

The carbon in the coal consists essentially of two types which behave differently during conversion.[6] The carbon associated with the volatile content of the coal is highly reactive at temperatures of about 1400° to 1700°F. The fixed carbon or residual char is less reactive, requiring temperatures above about 2000°F for conversion. The term coke is applied to char that has fused into lumps of a marketable size and quality, a property that depends on the coal type.

The sulfur content of coal in the United States ranges from about 0.3 to 8 percent.[4] Sulfur in coal is principally found in the form of either pyritic or organic sulfur. In the low sulfur coals, mainly the western coals, most of the sulfur is in the organic form, and in the high sulfur eastern coals most is in the pyritic form. Pyritic sulfur is sulfur combined with iron, mainly as pyrite but also as marcasite. These two minerals have the same chemical composition but different crystalline forms. Organic sulfur, on the other hand, is sulfur that is chemically linked to the coal. One distinction between the two forms is that the pyritic sulfur has a high specific gravity and can be separated from the coal by various physical means, of which the most important one is washing with water. Organic sulfur cannot be separated out physically and chemical means are needed, making it quite a difficult procedure.

Another important property of coal for synthetic fuel conversion is its agglomerating characteristic. Coals that agglomerate can cause blockages in the reactor. In this regard, high volatile coals are not necessarily ideal for conversion processes, as they tend to agglomerate at high temperature and pressure. There are other specific properties of coal that are of interest, including ash fusability, friability, hardness, and specific gravity; they will be discussed in context.

Shale

The shared property of all oil shales is the presence of "kerogen," a high molecular weight organic material which is almost totally insoluble in all common organic solvents. Oil shale sometimes contains a certain portion of soluble organic matter called soluble bitumen, but the major part of the pyrolysis oil is derived from the kerogen. Because the kerogen is not bound to a particular type of rock such as shale, the name oil shale is somewhat misleading. However, the largest concentrations of kerogen are found in sedimentary non-reservoir rocks such as marlstone.[7]

By definition, oil shale yields a minimum of 10 gallons of oil per ton of shale (written 10 gal/ton), and high grade shale greater than 25 gal/ton. From data presented in Ref. 3, we may derive the following empirical relation between the weight percent of total organic matter and the Fischer assay oil yield in gal/ton:

$$\text{Yield(gal/ton)} = 1.97 \times \text{Organic Matter(wt.\%)} - 2.59. \qquad (3.3)$$

This corresponds to about a 25 gal/ton yield for shale with 14 percent organic matter and 75 gal/ton for shale with 40 percent organic matter. The Fischer assay refers to the pyrolysis yield from a known weight of sample, under a controlled rate of heating to a temperature of 500°C (932°F) in the absence of air.

Table 3.2 shows a typical ultimate analysis of organic matter in the oil shale from the Mahogany Zone in the Piceance Creek Basin of Colorado.[3] A heating value may be assigned to shale using Eqs. (3.1) or (3.2), although oil shale is normally not combustible in its raw state. The higher heating value of the organic matter for the analysis in Table 3.2 is about 17,600 Btu/lb. The net value for a given shale is specified once the grade, that is, the fraction of organic matter, is

Table 3.2
Typical composition of organic matter
in Mahogany Zone shale.

Component	Weight Percent
Carbon	80.5
Hydrogen	10.3
Nitrogen	2.4
Sulfur	1.0
Oxygen	5.8
Total	100

known, since the inorganic mineral matter does not contribute to the heating value.

About half of the inorganic mineral portion of Green River shale consists of carbonates, of which about two thirds is dolomite and the remainder calcite. Feldspars make up about 20 percent of the inorganic material, and quartz and clays about 10 to 15 percent each. A good summary of the properties of oil shale may be found in Ref. 3.

3.3 Pyrolysis

When coal or shale is heated, or pyrolyzed, volatile material is distilled. The volatile material consists of condensable tar, oil and water vapor, and noncondensable gases. The carbon and mineral matter remaining behind is the residual char. Pyrolysis is one method to produce liquid fuels from coal, and it is the principal method used to convert oil shale to liquid fuels (see Fig. 3-1). Moreover, as gasification and liquefaction are carried out at elevated temperatures, pyrolysis may be considered a first stage in any conversion process. It is for this reason that we begin our detailed discussion of the fundamentals of conversion with this topic.

The Fischer assay pyrolysis yields from shale at 932°F (500°C) have been given in the previous section. Typical Fischer assay yields for various rank "as-received" coals[3] are shown in Table 3.3. At temperatures above about 1400°F the liquid yields rapidly drop off.

The pyrolysis tars and oils are not suitable final fuel products. They are solids at room temperature, becoming fluids at temperatures of about 100° to 130°F. Often they are also unstable and, when warmed, they polymerize and become

Table 3.3
Typical Fischer assay yields from different rank coals.

	Char (wt. %)	Tar (gal/ton)	Light Oil (gal/ton)	Gas (ft³/ton)	Water (wt. %)
Med. vol. bituminous	83	19	1.7	1,900	4
High vol. B bituminous	70	30	2.2	2,000	11
Subbituminous A	59	21	1.7	2,700	23
Lignite A	37	15	1.2	2,100	44

more viscous. Ash and mineral matter are removed in pyrolysis, which increases the heating value, but sulfur and nitrogen are not completely removed. A more stable and useful product is obtained by hydrogenating and by removing the sulfur and nitrogen in the fuel as hydrogen sulfide and ammonia. These procedures are, as noted previously, similar to the various known refinery procedures used to upgrade natural crude oils.

The composition of the noncondensable gases depends on whether the pyrolysis vessel or retort is directly or indirectly heated. In the indirectly heated processes a separate furnace is used to heat a solid or gaseous heat carrier, which is then mixed with the fuel to provide the process heat. When the heat carrier is an inorganic solid, like ceramic or sand particles, then the method yields a gas which is a true product of the coal or shale. When heat for pyrolysis is supplied directly by air- or oxygen-blowing to burn a portion of the coal, and if steam is introduced to control the temperature, the resultant gases are not true pyrolysis products but may be similar to gases produced by gasification. In either the indirect case or the direct oxygen-blown case, the heating value of the gases may be increased by the various processes discussed in the next section to make a pipeline gas. Alternatively, the gases may be used on-site for fuel or power generation.

The water vapor which distills off during pyrolysis is condensed and separated from the oil. This water, which we have called process condensate in the case of shale pyrolysis, can amount to as much as 10 gallons per ton of shale,[3] or typically from 30 to 40 percent by weight of the amount of oil produced. For coal the quantities of water per unit weight of raw material are much more, averaging about 25 gallons per ton for the higher yield bituminous coals or about 80 percent by weight of the amount of tar and oil produced (see Table 3.3). In either case the water is very dirty and contains dissolved minerals and organic materials, generally including phenols, organic acids, ammonia, and a variety of salts.

The char remaining after coal has been pyrolyzed has an improved heating value, when compared with the orginal coal, as a result of the removal of moisture. Sulfur removal, however, is not significant. In plants designed for fuel production the char is usually gasified, although low sulfur chars may be used directly as fuel. Spent shale is primarily inorganic residue but may contain some carbonaceous matter, and for that reason we do not refer to it as shale ash. Oil shales retorted without combustion of residual carbon in the ash may contain from 2 to 5 percent residual organic carbon.[7] More importantly, 80 to 85 percent by weight of the originally mined oil shale remains as spent material to be discarded. This feature still looms as the primary impediment to the above ground retorting of this valuable hydrocarbon resource.

3.4 Gasification of Coal

The different routes by which low-, medium-, and high-Btu gas can be produced from coal are shown in Fig. 3-1. Product fuel gas of low heating value (100-250 Btu/scf) is a consequence of blowing the gasifier with air. This leads to large concentrations of inert nitrogen difficult to separate from the fuel gas. If the gasifier is blown with oxygen instead of air, this problem is eliminated and a medium-Btu gas with a heating value between 250 and 400 Btu/scf results. A medium-Btu gas of higher heating value can be obtained by hydrogasification— by blowing the gasifier with a hydrogen-steam mixture instead of oxygen and steam—thereby enhancing the rate of production of methane.

High-Btu or pipeline gas is composed of more than 90 percent methane, which has a heating value of about 1000 Btu/scf, the remainder being mainly hydrogen and carbon dioxide. The heating value of pipeline gas is generally between 920 and 1000 Btu/scf. An objective in the production of pipeline gas is to produce as much methane (CH_4) as possible directly in the gasifier and so reduce the amount of subsequent upgrading, termed methanation (see Fig. 3-1). The operating conditions most favorable for methane formation are suggested by a study of the several chemical reactions occurring in a gasifier.

The initial stage of gasification is one of pyrolysis. As the coal is heated, moisture is rapidly driven off and in countercurrent processes is immediately carried away in the exiting gas stream. At high temperatures, carbon oxides and hydrocarbons (mainly methane) are released, so the pyrolysis stage may be represented by:

Pyrolysis:

$$\text{Heat} + \text{Coal} \rightarrow CO, \; CO_2, \; CH_4. \qquad (3.4)$$

The subsequent gasification chemistry is complex but for the purposes of this discussion may be represented by the following five reactions:

Combustion ($\frac{1}{2} \leqslant n \leqslant 1$):

$$C + nO_2 \rightarrow (2 - 2n)CO + (2n - 1)CO_2 \quad \Delta H = +(2 - 2n)47.5$$
$$+ (2n - 1)169 \quad (3.5)$$

Carbon-steam or gasification reaction:

$$C + H_2O \text{ (steam)} \rightarrow CO + H_2 \qquad\qquad \Delta H = -56.5 \qquad\qquad (3.6)$$

Carbon-hydrogen or hydrogenation reaction:

$$C + 2H_2 \rightarrow CH_4 \qquad\qquad\qquad \Delta H = +32.2 \qquad\qquad (3.7)$$

Water-gas shift reaction:

$$CO + H_2O \text{ (steam)} \rightleftharpoons H_2 + CO_2 \qquad \Delta H = +17.7 \qquad\qquad (3.8)$$

Methanation reaction:

$$CO + 3H_2 \rightarrow CH_4 + H_2O \qquad\qquad \Delta H = +88.7 \qquad\qquad (3.9)$$

Here, ΔH is the heat of reaction in 10^3 Btu released per lb molecular weight (mole) of C or CO reacted, when the reaction proceeds from left to right.

The ash in the coal is untouched by the gasification process and is removed either as a solid (dry ash) or as a melt (slag). The nitrogen and sulfur are converted to ammonia (NH_3) and hydrogen sulfide (H_2S) and can be removed by various proven processes.

The gasification reaction (3.6) is endothermic. This means that it will not take place without heat, which is usually supplied by burning the coal or char as represented by the reactions (3.5). The gas composition leaving the gasifier is a function of the relative contribution of each reaction to the overall process and thus is determined by the rate of the individual reactions and by the residence time of reactants and products in the gasifier. Only the combustion reaction (3.5) goes to completion; oxygen does not leave the gasifier. Reaction rates may be increased by increasing the proportion of specific components in the feed. For example, increasing the amount of steam increases the carbon-steam reaction rate.

Operating conditions of temperature and pressure also affect the relative reaction rates. At low temperatures the value of n in Eq. (3.5) is close to one so that mainly CO_2 is produced with considerable heat released.[8] As the temperature increases, n approaches a value of $\frac{1}{2}$, and thus CO is preferentially formed with less heat released. Increasing the proportion of steam in the feed reduces operating temperatures because the carbon-steam reaction is endothermic.[9] High temperature operation at say 3000°F, in addition to producing a relatively low CO_2 gas, also ensures conversion of a high fraction of the coal, because the char as well as the highly reactive carbon is gasified. Low temperature operation may sometimes be desirable; these conditions favor the carbon-hydrogen reaction (3.7), which increases the amount of methane formed in the gasifier.

In hydrogasification, the rate of methane production is increased by blowing the gasifier with hydrogen and steam instead of oxygen and steam.[9] Some hydrogasifiers have a second stage operating at a high temperature to increase the amount of coal converted.[10] A further advantage of a high temperature second stage is its favorable conditions for the carbon-steam reaction (3.6), which produces "free" hydrogen that can be used for hydrogenation in the first stage. The unconverted char is used to produce the main hydrogen feed from water.

The raw gases from both oxygen-blown and hydrogen-blown gasifiers fall in the medium-Btu category and require further treatment if pipeline gas is the desired product. A simplified flow sheet for pipeline gas manufacture is shown in Fig. 3-3. The gas purification stage, that is, removal of the sulfur gases and CO_2, is discussed in Section 3.6. The conversion of the remaining constituents to

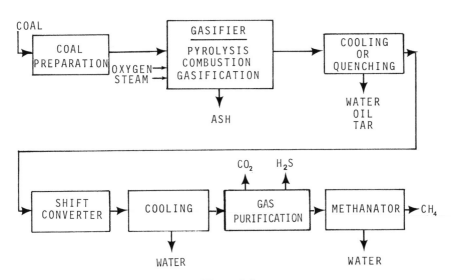

Figure 3-3
Stages in the production of pipeline gas using an oxygen-blown gasifier.

methane is in two stages. First, steam is introduced into the gas in a "shift" converter that contains iron-chromium oxide pellets or other materials suitable for catalyzing the water shift reaction (3.8). In this process the ratio of H_2 to CO is increased to 3-to-1, as next required for the methanation stage in which the hydrogen and carbon monoxide react in the presence of a nickel-based or other catalyst to form methane, following the reaction (3.9). Water of good quality is ejected from the methanator.

All three principal reaction units in pipeline gas production—the gasifier, shift converter, and methanator—release heat. Estimates of the heat released in these units are given in Table 3.4. These values were calculated using the reactions shown and assuming the feed materials to be in the correct proportions to yield the indicated reaction products. Theoretically, the heat released can be recovered and fully utilized for preheating the coal and for producing the required feed materials—the steam and the oxygen or hydrogen. In practice, however, some fraction of the heat relased by each unit is rejected to the environment. It is desirable then to limit the exothermicity of the overall process as much as possible.

Table 3.4
Calculated heat release in principal reaction units
based on a simplified gasification scheme.

Reactions	Feed (lb moles @ 25°C)	Product (lb moles @ 25°C)	Heat Released (10^3 Btu)
In Gasifier			
Combustion,	C	aCH$_4$	$-33a$
Carbon-steam,	H$_2$O	bH$_2$	$+104b$
Carbon-hydrogen	O$_2$	cCO	$+48c$
	zH$_2$	dCO$_2$	$+169d$
		(no water)	$-9z$
In Shift Converter			
Water-gas shift	aCH$_4$	aCH$_4$	$4.5(3c - b)$
	bH$_2$	$\frac{3}{4}(b + c)$H$_2$	
	cCO	$\frac{1}{4}(b + c)$CO	
	dCO$_2$	$(d + \frac{3}{4}c - \frac{1}{4}b)CO_2$	
	$(\frac{3}{4}c - \frac{1}{4}b)H_2$O		
In Methanator			
Methanation	aCH$_4$	$[a + \frac{1}{4}(b + c)]$CH$_4$	$22.2(b + c)$
	$\frac{3}{4}(b + c)$H$_2$	$\frac{1}{4}(b + c)$H$_2$O	
	$\frac{1}{4}(b + c)$CO		

A study of Table 3.4 suggests that increasing the amount of hydrogen in the feed, and hence the relative amount of methane formed in the gasifier, will decrease the heat released in this unit. This explains the increasing interest in hydrogasification. Further, hydrogen-blown gasifiers can usually be operated to produce a raw gas with a H_2 to CO ratio close to 3-to-1, thereby eliminating the need for shift conversion and its associated heat loss.

In the gasification procedures described, the heat for the gasification reaction of carbon and steam is supplied directly by combustion, but like shale pyrolysis the heat can be supplied indirectly by a heat carrier. The carrier could be any one of a number of inert materials heated in a separate furnace. One material used is agglomerated ash from coal burnt in a separate furnace.[2] A carrier of chemical heat can also be used. An example of this is the CO_2 Acceptor process,[11] in which the gasification heat is supplied by the exothermic reaction of dolomite (CaO), termed the acceptor, with CO_2. The spent dolomite ($CaCO_3$) is removed and regenerated in a separate furnace whose heat reverses the acceptor reaction. In both examples of indirect heating, gasification is carried out in the reactor by blowing in steam alone. Since combustion takes place in a separate furnace, air and not oxygen can be used, and the gasifier product is a medium-, not a low-, Btu gas.

In summary then, the feed and operating conditions determine the product composition and the amount of heat released from the gasifier. High temperature operation ensures conversion of a high fraction of the coal and high CO to CO_2 ratios. Low temperature operation favors methane formation, which is desirable in the production of pipeline gas. The unconverted char residue from low temperature gasifiers may be used for producing hydrogen. Hydrogen-blown gasifiers yield relatively high methane concentrations and normally do not require shift conversion prior to methanation of the remaining gases. The associated reduction in heat losses has to be greater than heat losses incurred in the production of the hydrogen if the overall efficiency of the process is to be improved.

3.5 Hydrogenation of Coal and Pyrolysis Oils

Hydrogenation is the chemical combination of a substance with hydrogen, usually in the presence of a catalyst. Hydrogenation of coal and pyrolysis oils produces more volatile products; the greater the degree of hydrogenation the more volatile is the final product. It is the principal route for the production of clean solid and liquid fuels (see Fig. 3-1).

Solvent refined coal, produced in the presence of a hydrogen donor solvent, is an example of a material with a low degree of hydrogenation. Hydrogen gas reacts with the solvent and the reduced solvent, then gives up its hydrogen to the coal. The coal dissolves in the solvent and in this way is separated from the ash. Most of the hydrogen is consumed in reducing the oxygen, sulfur, and nitrogen following the reactions

$$O + H_2 \rightarrow H_2O, \tag{3.10}$$

$$S + H_2 \rightarrow H_2S, \tag{3.11}$$

$$N + \tfrac{3}{2}H_2 \rightarrow NH_3. \tag{3.12}$$

The water, hydrogen sulfide, and ammonia that are formed can be separated from the main product.

A higher degree of hydrogenation of coal yields oils. In the Synthoil process,[12] coal is slurried in a recycled product oil, mixed with hydrogen, and passed up a fixed bed of catalyst pellets at moderate temperatures around 850°F and high pressures from 2,000 to 4,000 psig. The process is quite effective in removing almost all of the oxygen and most of the sulfur. The H-Coal process[13] uses a different means to contact hydrogen and slurried coal with the catalyst, but the reactor operating conditions and conversion of coal are similar to those in the Synthoil process. Both processes can give lighter hydrocarbons by using more hydrogen.

Hydrogenation of oils obtained by pyrolysis of coal or shale is used to reduce the sulfur and nitrogen levels and to produce a pumpable synthetic crude oil suitable as a refinery feedstock.

The last method by which clean liquid fuels may be derived from coal is by Fischer-Tropsch synthesis. In the synthesis process carbon monoxide and hydrogen are reacted in the presence of a catalyst to form hydrocarbon vapors, and these are condensed to the liquid fuels. It may be recalled that in the production of high-Btu gas, part of the methane is formed by reacting carbon monoxide and hydrogen in the presence of a catalyst and under temperature and pressure conditions specific to the methanation reaction (3.9). Other catalysts are used when different hydrogenated products are required. Hydrocarbon compounds whose general formula is C_nH_{2n+2} (aliphatic hydrocarbons) can be synthesized, as well as compounds of carbon, hydrogen, and oxygen, such as methanol (oxyhydrocarbons). The reaction products vary according to the characteristics and type of catalysts and promoters used. The syntheses of gasoline, kerosene, diesel fuels, fuel oil, and other aliphatic hydrocarbons are carried out typically with iron, cobalt, and nickel catalysts, and may be represented by the following general reactions:

Fischer-Tropsch reactions:

$$n\text{CO} + (2n + 1)\text{H}_2 \rightarrow \text{C}_n\text{H}_{2n+2} + n\text{H}_2\text{O}, \qquad (3.13)$$

$$n\text{CO} + 2n\text{H}_2 \rightarrow \text{C}_n\text{H}_{2n} + n\text{H}_2\text{O}. \qquad (3.14)$$

The catalytic conversion of carbon monoxide to methanol may be represented by the following reaction:

Methanol reaction:

$$\text{CO} + 2\text{H}_2 \rightarrow \text{CH}_3\text{OH}. \qquad (3.15)$$

Water is the source of the hydrogen for all of the hydrogenation processes but its reduction to hydrogen may be carried out in a variety of ways, either directly or indirectly. Hydrogen is directly produced from water by reduction with carbon (char), following the reaction

$$\text{Heat} + \text{H}_2\text{O} + \text{C} \rightarrow \text{CO} + \text{H}_2. \qquad (3.16)$$

The carbon monoxide is then made to undergo the water-gas shift conversion of Eq. (3.8), and the CO_2 that is formed is removed by use of gas purification procedures discussed in the next section. The reduction of water to hydrogen may also be indirect, using iron to reduce the water to hydrogen and oxides of iron.

The theoretical minimum heat required to produce hydrogen from steam is about 320 Btu/scf. This heat may be supplied by the combustion of excess char or may be supplied indirectly, for example, using electrical energy. In either case this heat forms part of the total heat input to the process and must be accounted for when determining the overall conversion efficiency of any process.

3.6 Gas Purification

Gas leaving a gasifier or other reactor usually contains components that make it unsuitable for direct use or further processing. The undesirable components are removed as the gas passes through the treatments shown in Fig. 3-3. The gas may contain water vapor, oil, tar, and particles of char or ash. When the gas is cooled, oil and water condense, bringing with them tar and solid particles. A separation vessel is used to cause the cooled stream to split into four streams—gas, oil, water, and tar and solids.

Ammonia, formed by the hydrogenation of nitrogen in the coal or shale, is very soluble and is removed nearly completely from the gas stream by absorption in the condensed water. However, cooling and scrubbing is seldom sufficient to adequately purify the gas. Sulfur must usually be removed to render the gas fit for combustion without air pollution or to protect methanation catalysts. If a high-Btu gas is the desired product, then carbon dioxide must be removed. Removal of the acid gases, carbon dioxide, and hydrogen sulfide, takes place in a separate gas purification step (see Fig. 3-3).

In all the currently practiced, large-scale, acid gas removal procedures,[14-16] the acid gases are selectively dissolved in a liquid, passed countercurrent to the gas. In a separate vessel the absorbing liquid is stripped of its gas content and thereby regenerated. It is then recycled back to the absorber. A simple flow scheme is shown in Fig. 3-4. Absorption is caused by any or all of low temperature, high pressure, or the use of an alkaline solution. Regeneration, or desorption, is caused by reduced pressure, high temperature, and boiling. This causes the solvent vapor to rise through the liquid, stripping out the acid gases.

Many synthetic natural gas plants operate at high pressures, up to 1,000 psig. At these pressures, simple dissolution of the acid gases in an organic solvent is practical. The processes are mostly proprietary and are known by trade names such as Rectisol, Selexol, Sulfinol, Fluor, and Purisol. They share the feature that absorption of the acid gas takes place at temperatures generally below 120°F. Thus the gas and circulated solvent must be strongly cooled with cooling water and refrigeration is often required. In all the processes the solvent must be boiled to regenerate it.

A commonly used alkaline solution is a 20 to 30 percent solution of potassium carbonate in water. Absorption and regeneration take place at about the same temperature, in the range of 180° to 300°F, so that strong cooling is not required. Absorption requires pressures above about 300 psig. In regeneration, the pressure is released and the solution boiled. The steam used to boil the solvent is usually directly injected into the solvent or has its heat transferred through a heat transfer surface.

Water condenses from the gas streams when they are cooled before entering the absorber. Gas purification, however, neither consumes nor releases water if the gas enters and leaves the gas purification system at the same pressure and temperature. If direct steam injection is used, the recovered water is dirty and

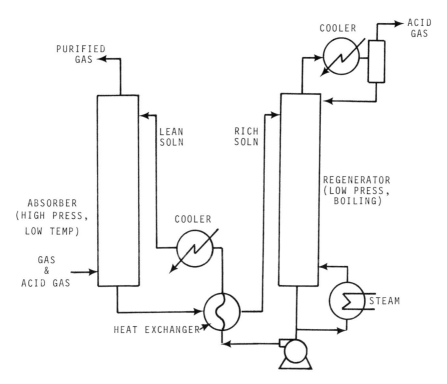

Figure 3-4
The simplest gas purification arrangement.

requires cleaning before it is recycled back to a boiler. For the purposes of this book, the importance of gas purification lies in its large consumption of energy and the ultimate disposal of this energy. Between 5.5 and 6 percent of the heating value in the coal fed to a pipeline gas plant is consumed in gas purification.

Most of the energy for gas purification goes into boiling solvent, which all the regeneration schemes involve, and this energy is ultimately dissipated by condensing the solvent. If the boiling point of the solvent at the pressure in the regenerator is high, the condenser can lose its heat directly to the atmosphere without the use of cooling water. Although an air cooled condenser is more expensive than a water cooled condenser, air cooling is frequently used in arid regions. With reasonable water conservation planned, air cooling is the method of choice with potassium carbonate as the solvent. For organic solvents, about 90 percent of the cooling needs can best be met by air cooling, with the remainder met by water cooling.

Table 3.5
Energy requirements per lb mole of acid gas removed
in gas purification systems.

Solvent	Regenerator Steam (Btu)	Solvent Pump (kwh)
Potassium Carbonate	30,000	0.26
Organic	28,400	0.30

The true energy and cooling requirements of gas purification systems, particularly those employing organic solvents, are for the most part proprietary and partly controlled by the capital investment made. For estimating purposes Table 3.5 lists the principal energy requirements in plants where reasonable water conservation is planned.

3.7 Flue Gas Desulfurization

In some synthetic fuel plants, coal or char will be burnt for the purpose of generating steam for process needs or to produce electric power. The burning of char and coal, even low sulfur western coal, generates both particulates of fly ash and oxides of sulfur in the exiting flue gases. Particulate removal to meet air quality standards need not involve any significant water consumption if electrostatic precipitators are used, although some water may be consumed in disposing of the fly ash. Even when wet scrubbing techniques are used to remove the fly ash, this is generally done in conjunction with sulfur removal. Therefore the particulate scrubbing may be looked upon as a beneficial side effect, not appreciably adding to the water consumption beyond that needed to remove the sulfur. The water consumed in either the dry or wet methods of particulate removal is mostly associated with the fly ash disposal itself and this will be examined in Chapter 7 in the context of solids disposal.

Sulfur dioxide (SO_2) represents 98 percent of the sulfur oxide pollutants, and its removal from the combustion gases before they are released to the stack is an important use of water in most procedures. A large number of removal procedures exist or are under development. They may be classified as wet or dry, depending on whether a water mixture is used to absorb the SO_2 or whether the acceptor is dry. The procedures are further classified as non-regenerative or regenerative, depending on whether the chemical used to remove the SO_2 is disposed of or regenerated and reused.

Dry processes will not be considered further. None have been fully tested nor do they involve the direct use of water. Nevertheless, the indirect use of water even in such procedures must be borne in mind. For example, in the potentially promising Shell copper oxide process,[17] SO_2 is absorbed by copper oxide at temperatures of around 750°F and hydrogen, most likely produced from water, is used for regeneration.

In wet non-regenerative procedures, a sludge is produced which is partially dewatered and then disposed. The water remaining in the sludge constitutes a consumption. In regenerative processes either elemental sulfur or sulfuric acid is normally recovered. This recovery generally entails an indirect use of water. Most present and projected installations are of the non-regenerative type.

Wet scrubbing processes contact the flue gas with a solution or slurry into which the SO_2 is absorbed. The most likely procedures for use in the near future are the non-regenerative systems that utilize limestone ($CaCO_3$), hydrated or slaked lime ($Ca(OH)_2$), or both. In the wet limestone process the flue gas contacts a wet limestone slurry in the scrubber and the SO_2 is removed following the reaction

$$CaCO_3 + SO_2 \rightarrow CaSO_3 + CO_2, \qquad (3.17)$$

with

$$CaSO_3 + \tfrac{1}{2}O_2 \rightarrow CaSO_4. \qquad (3.18)$$

The scrubber effluent generates a spent slurry following the reactions

$$CaSO_3 + \tfrac{1}{2}H_2O \rightarrow CaSO_3 \cdot \tfrac{1}{2}H_2O\downarrow, \qquad (3.19)$$

$$CaSO_4 + 2H_2O \rightarrow CaSO_4 \cdot 2H_2O\downarrow. \qquad (3.20)$$

In the wet lime scrubbing system the gas containing the SO_2 is reacted with a wet lime slurry following the reaction

$$Ca(OH)_2 + SO_2 \rightarrow CaSO_3 + H_2O. \qquad (3.21)$$

Reactions (3.18) to (3.20) follow.

In the so-called double alkali, non-regenerative procedure[18] the flue gas is scrubbed with a soluble alkali such as sodium sulfite, which is subsequently regenerated with an insoluble alkali such as lime. Typical reactions in the scrubber and reaction tank are, respectively,

$$Na_2SO_3 + SO_2 + H_2O \rightarrow 2NaHSO_3, \qquad (3.22)$$

$$2NaHSO_3 + Ca(OH)_2 \rightarrow Na_2SO_3 + 2H_2O + CaSO_3. \qquad (3.23)$$

The spent calcium sulfite slurry must be disposed of, presenting similar problems to the lime/limestone procedures. The Wellman-Lord process[19] is an example of a wet regenerative procedure in which absorption is by sodium sulfite, as represented by reaction (3.22). The spent absorbent, sodium bisulfite, is regenerated thermally following the reaction

$$2NaHSO_3 \xrightarrow{\text{heat}} Na_2SO_3 + SO_2 + H_2O. \tag{3.24}$$

The water vapor is condensed and the active sodium sulfite dissolved in it for recycling to the scrubber. The concentrated SO_2 stream can be processed to elemental sulfur or sulfuric acid in a suitable plant.

The important feature in all of these procedures is that water leaves the system as vapor in the flue gas. Additionally, in the non-regenerative processes, water also leaves in the slurry of spent solids. Methods for estimating the water in these two streams, which are independent of the specific scrubbing procedure or its details, are presented below.[20] They can therefore be applied to any wet system.

To estimate the quantity of water that leaves in the scrubbed flue gas, the gas is assumed to be saturated with water vapor at the temperature and pressure at which it leaves the scrubber, in the final absorber and before any reheating. Reheating is usually necessary to provide sufficient buoyancy to the stack gases cooled down in the scrubbing procedure. The dependence on temperature and pressure of the water content is illustrated in Table 3.6. From the table it is clear that lack of knowledge of the pressure of saturation in the range of interest will give an error of not more than 7 percent. If, however, the temperature of saturation is 10°F higher than was assumed, the water content of the flue gas will be 35 to 40 percent higher than calculated. This indicates a severe limitation on the ability to estimate the quantity of water in the flue gas without detailed and precise information on the gas temperature.

Table 3.6
Moles of water vapor per mole of dry gas at saturation.

Temperature (°F)	Pressure (inches of water gauge)		
	0	10	20
120	0.130	0.127	0.123
130	0.178	0.173	0.168
140	0.245	0.237	0.231
150	0.339	0.328	0.318

The dry flue gas volume must also be known. Assuming negligible carbon monoxide and nitrogen oxides, the flue gas components are taken to be carbon dioxide, water vapor, sulfur dioxide, oxygen, and nitrogen. The total moles of dry flue gas per unit weight of coal or char, as fired, is given by[20]

$$\frac{\text{moles of dry flue gas}}{\text{lb of coal or char}} = (4.76)\,(1 + a)\left(\frac{c}{12} + \frac{s}{32}\right) + (3.76 + 4.76a)\left(\frac{h}{4} - \frac{x}{32}\right). \quad (3.25)$$

Here, c, s, h, and x are the weights of carbon, sulfur, hydrogen, and oxygen per unit weight of coal or char, and a is the fraction of excess air to the burner. The water carried away by the gas, which was not in the gas before scrubbing (the makeup water requirement for gas saturation), then becomes

$$\frac{\text{lb makeup water}}{\text{lb coal or char}} = \left(\frac{\text{moles of dry flue gas}}{\text{lb of coal or char}}\right)\left(\frac{\text{moles of water vapor}}{\text{moles of dry flue gas}}\right)(18) - w - \frac{h}{9}, \quad (3.26)$$

where w is the fractional weight of moisture in the coal. For the char, w is zero. If knowledge of the scrubber equilibrium temperature and pressure is not available, a value of 0.127 is suggested for the moles of water vapor per mole of dry flue gas. This value has been found to give results close to the few available experimental results recorded.

The excess air fraction, a, varies from about 0.05 to 0.2. A value of $a = 0.15$ may be used for estimations, recognizing that changes in this quantity over the range indicated will generally not introduce an error greater than 10 percent. For high moisture content coals such as lignite where w is large, the relative error would be somewhat larger because the absolute makeup requirement is small. The moisture content of the coal is the largest single factor affecting the makeup water requirement. Comparison of the water requirements for different coals calculated by Eq. (3.26) with more detailed estimates is quite favorable.[20]

In either the lime/limestone or sodium-lime scrubbing processes, the two major solid wastes are $CaSO_3 \cdot \frac{1}{2}H_2O$ or $CaSO_4 \cdot 2H_2O$. In addition, unreacted limestone, $CaCO_3$, and slaked lime, $Ca(OH)_2$, are also wasted. The amount of water leaving in the waste solids depends on the quantity of sulfur and the slurry concentration. From data on the composition of lime and limestone sludges,[21] the weight of solids and the water of hydration per unit weight of sulfur have been calculated for the lime and limestone processes.[20] These figures are given in Table 3.7. Also shown in this table are the comparable figures for the crystalline forms of calcium sulfate and calcium sulfite.

Table 3.7
Weight of solids and water of hydration per unit weight
of sulfur in lime and limestone sludges, and for crystalline
forms of calcium sulfate and sulfite.

Crystal or Process	lb solid/lb sulfur	lb water/lb sulfur
$CaSO_3 \cdot \frac{1}{2}H_2O$	4.0	0.28
$CaSO_4 \cdot 2H_2O$	5.4	1.13
Lime	5.2	0.38
Limestone	6.6	0.38

Table 3.7 shows that the water of hydration in the lime and limestone processes can represent only a very small fraction of the total makeup water (slurry
water plus water of hydration). The contribution of the hydration water may
therefore be neglected and the makeup water be taken to equal the slurry water.
In this case we write

$$\frac{\text{lb makeup water}}{\text{lb sulfur}} = \left(\frac{\text{lb solid}}{\text{lb sulfur}}\right)\left(\frac{1 - m}{m}\right), \qquad (3.27)$$

where m is the weight fraction of solids in the waste (the weight of solids divided
by the weight of water plus solids). Note that a change to 30 percent solids from
40 percent solids makes a 50 percent increase in the water in the waste, since
$(1 - m)/m$ changes from 1.5 to 2.3. A change to 50 percent solids makes
$(1 - m)/m$ equal to 1 and decreases the water in the waste by 33 percent. The exact
value of the percent solids will of course depend upon the dewatering method
used. For example, a 40 percent solids concentration represents a well dewatered
waste obtained by gravity thickening. Vacuum filtration could increase the percent solids to 60 percent and with centrifugation added could raise this figure to
70 percent.[21]

The weights of solids per unit weight of sulfur for the lime and limestone
processes are given in Table 3.7, with an average of 5.9 lb solid/lb sulfur
appropriate for a mixed process. For the sodium-lime processes represented by
Eq. (3.23), a fractional weight of about 5 lb solid/lb sulfur should provide a
satisfactory estimate. Comparison of the slurry water requirements calculated by
Eq. (3.27) agree well with design estimates from actual scrubber systems.[20]

Chapter 3. References

1. Bodle, W. W., Vyas, K. C., and Talwalkar, A. T., "Clean Fuels from Coal, Technical-

Historical Background and Principles of Modern Technology," *Clean Fuels from Coal Symposium II,* pp. 53–84, Institute of Gas Technology, Chicago, Ill., 1975.

2. Tetra Tech, Inc., "Energy From Coal—A State-of-the-Art Review," ERDA Report No. 76-7, U.S. Gov't. Printing Office, Washington, D.C., 1976.

3. Hendrickson, T. A., *Synthetic Fuels Data Handbook.* Cameron Engineers, Inc., Denver, Colo., 1975.

4. Averitt, P., "Coal Resources of the United States, January 1, 1974," Geological Survey Bulletin No. 1412, U.S. Gov't. Printing Office, Washington, D.C., 1975.

5. Perry, R. H. and Chilton, C. H. (eds.), *Chemical Engineers' Handbook,* Section 9. 5th Edition, McGraw-Hill, New York, 1973.

6. Wen, C. Y. and Huebler, J., "Kinetic Study of Coal-Char Hydrogasification," *Ind. Eng. Chem. Process Design & Develop.* **4,** 142–154 (1965).

7. Schmidt-Collerus, J. J., "The Disposal and Environmental Effects of Carbonaceous Solid Wastes from Commercial Oil Shale Operations," Report No. NSF-RA-E-74-004 (NTIS Catalog No. PB 231 796), Denver Research Institute, Denver, Colo., Jan. 1974.

8. Wen, C. Y., "Optimization of Coal Gasification Processes," R&D Report No. 60, Interim Report No. 1, Office of Coal Research, Washington, D.C., 1972.

9. Feldkirchner, H. L. and Huebler, J., "Reaction of Coal with Steam-Hydrogen Mixtures at High Temperatures and Pressures," *Ind. Eng. Chem. Process Design & Develop.* **4,** 132–142 (1965).

10. Lee, B. S., Pyrcioch, E. J., and Schora, F. C., "Hydrogasification of Pretreated Coal for Pipeline Gas Production," *Advances in Chemistry Series No. 69,* American Chemical Society, Washington, D.C., 1967.

11. Fink, C., Curran, G., and Sudbury, J., "CO_2 Acceptor Process Pilot Plant—1974, Rapid City, South Dakota," *Clean Fuels from Coal Symposium II,* pp. 243–257, Institute of Gas Technology, Chicago, Ill., 1975.

12. Friedman, S., Yavorsky, P. M., and Akhtar, S., "The Synthoil Process," *Clean Fuels from Coal Symposium II,* pp. 481–494, Institute of Gas Technology, Chicago, Ill., 1975.

13. Johnson, C. A. *et al.,* "Present Status of the H-Coal Process," *Clean Fuels from Coal Symposium II,* pp. 525–551, Institute of Gas Technology, Chicago, Ill., 1975.

14. Riesenfeld, F. C. and Kohl, A. L., *Gas Purification.* 2nd Edition, Gulf Publishing Co., Houston, Texas, 1974.

15. Maddox, R. H., *Gas and Liquid Sweetening.* 2nd Edition, John M. Campbell Co., Norman, Oklahoma, 1974.

16. Colton, C. B., Dandarati, M. S., and May, V. B., "Low and Intermediate Btu Fuel Gas Cleanup," *Symp. Proc: Environmental Aspects of Fuel Conversion Technology II,* pp. 193–215, Report No. EPA-600/2-76-149, Environmental Protection Agency, Research Triangle Park, N.C., June 1976.

17. Pohlenz, J. B., "The Shell Flue Gas Desulfurization Process," *Proc.: Symposium on Flue Gas Desulfurization, Atlanta, November 1974,* Vol. II, pp. 807–835, Report No.

EPA-650/2-74-126-b, Environmental Protection Agency, Research Triangle Park, N.C., Dec. 1974.

18. Kaplan, N., "Introduction to Double Alkali Flue Gas Desulfurization Technology," *Proc.: Symposium on Flue Gas Desulfurization, New Orleans, March 1976,* Vol. I, pp. 387–421, Report No. EPA-600/2-76-136a, Environmental Protection Agency, Research Triangle Park, N.C., May 1976.

19. Pedroso, R. I., "An Update of the Wellman-Lord Flue Gas Desulfurization Process," *Proc: Symposium on Flue Gas Desulfurization, New Orleans, March 1976,* Vol. II, pp. 719–733, Report No. EPA-600/2-76-136a, Environmental Protection Agency, Research Triangle Park, N.C. May 1976.

20. Gold, H., Goldstein, D. J., Probstein, R. F., Shen, J. S., and Yung, D., "Water Requirements for Steam-Electric Power Generation and Synthetic Fuel Plants in the Western United States," Report No. EPA-600/7-77-037, Environmental Protection Agency, Office of Energy, Minerals & Industry, Washington, D.C., April 1977.

21. Cooper, H. B., "The Ultimate Disposal of Ash and Other Solids from Electric Power Generation," *Water Management by the Electric Power Industry* (E. F. Gloyna, H. H. Woodson, and H. R. Drew, eds.), pp. 183–195, Center for Research in Water Resources, The University of Texas at Austin, 1975.

4
Cooling Fundamentals

4.1 Evaporative Cooling Towers

Water used for cooling passes through a heat exchanger where the heat to be dissipated is transferred to the water. The cooling water may pass through only once and be discharged after heating up or it may be cooled down and recirculated. A wet or evaporative cooling tower is one means of removing the heat from the water before it is recirculated. In this type of tower, shown schematically in Fig. 4-1, the warm water leaving the heat exchanger is pumped to the top, distributed through an assembly of nozzles, and allowed to fall through the tower in direct contact with the air that removes the heat. Packing material inside the tower retards and breaks up the water into small droplets, creating both a large surface area of liquid and a high residence time for the liquid in the tower. The transfer of heat from the warm droplets to the air is more rapid when the surface area of the droplets is larger. The heat is transferred by evaporation and by convection causing the air to rise in temperature (sensible heating).

The water that is evaporated and the heat that is lost are absorbed by the air moving through the tower. In a natural draft (hyperbolic) tower, air is drawn in by natural convection caused by the difference in densities between the inside and outside of the tower. In the mechanical draft tower, the air is drawn in by fans. To maintain a constant water circulation rate it is necessary to "makeup" the water lost through evaporation and drift and the water that is removed (blowdown) to eliminate the accumulation of solids.

Very large and often unrealistic estimates have appeared in the literature of the amount of water that must be evaporated to meet the cooling needs of synthetic fuel plants. The reason for this lies primarily in two assumptions: all of the unrecovered heat in the plant ends up evaporating water; and cooling towers evaporate water at the rate of 1 lb/10^3 Btu of heat transferred. As we mentioned earlier, the assumption concerning the heat load is not correct, since some of the unrecovered heat in the plant is lost directly to the atmosphere in a variety of ways, as in hot flue gases and in coal drying. These direct losses can amount to from 20 to 40 percent of the unrecovered heat. Moreover, even in water rich areas it is likely that an additional 20 to 40 percent of the unrecovered heat would be dissipated more economically by dry cooling. The second assumption concerning the evaporation rate may be seriously in error by failing to include the sensible heat transferred to the large mass of air passing through the tower and disregarding differences in evaporative heat transfer associated with varying ambient air conditions. The evaporation rate in a tower depends on several factors: the relative flow rates of air and water, the entering warm water tempera-

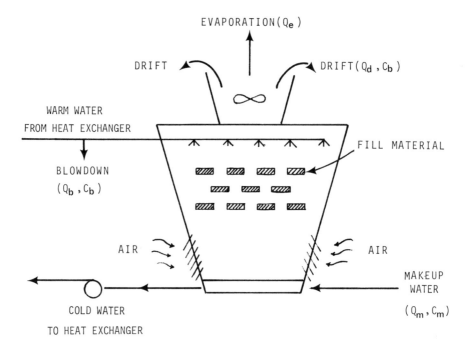

Figure 4-1
Wet cooling tower.

ture, and the temperature and humidity of the entering air, which determine the ability of the air to absorb water. Monthly and annual variations in the ambient air conditions are therefore important factors in assessing evaporation rates. The design of the tower and to some extent how it is operated are also important.

Leung and Moore[1] (see also, Ref. 2) have determined, for a fixed heat removal rate and fixed flow conditions, the rate of evaporative water loss in a mechanical draft tower as a function of the wet-bulb temperature and relative humidity of the ambient air. The wet-bulb temperature is very close to the adiabatic saturation temperature for air-water vapor mixtures and the lowest possible temperature to which the water can be cooled. Cooling to the wet-bulb temperature is, however, not a realistic limit since an infinitely tall cooling tower would be required. In their calculations, Leung and Moore took the difference between the temperature of the cold water leaving the tower and the wet-bulb temperature of the entering air to be 17°F. This temperature difference is termed the approach temperature. They chose the cooling range for the tower (the difference between the hot and cold water temperatures) to be 25°F and the ratio of the flow rates of circulating water to dry air mass to be 1.6. Together with the heat load and wet-bulb temperature these conditions fix the size of the tower for the all-wet cooling of steam-electric plant turbine condensers. For a given heat load, a lower temperature range would mean higher water flow rates and higher pumping costs with larger condenser sizes.[1] When the circulating water flow is fixed, a lower ratio of water to air flow would require greater air flow with higher fan power. In Fig. 4-2 results from Ref. 2 are shown with the curves extrapolated to somewhat lower and higher wet-bulb temperatures than given in the original reference. These results show that with a high wet-bulb temperature and low relative humidity, that is, when the air is hot and dry, the evaporative loss is high and the sensible heat transfer capability low. On the other hand, with a low wet-bulb temperature and high relative humidity, when the air is cold and damp, the opposite is true.

In a synthetic fuel plant the cooling load on steam turbine condensers may be a large part of the plant cooling. Most of this load would be wet cooled. The cooling tower design is therefore related to the characteristics of the steam turbine condensers, and there is a question as to whether the results of Fig. 4-2 are applicable, since they represent a design for steam-electric turbine condensers. The important difference is that the towers are designed with a fixed circulating water flow rate, and hence the condenser has a fixed maximum temperature when the ambient air is at summer conditions. But the performance of a generating plant depends directly on the temperature of the condensing steam, with a lower condensing temperature giving a higher efficiency. Therefore, when the ambient air temperature drops during the colder months of the year, the condenser temperature also drops, and the plant efficiency increases. In a synthetic fuel plant, however, the economics would probably dictate the bypassing of some of the water from the cooling tower without cooling it and thereby maintaining the cold water temperature at a fixed value rather than lowering it. Because the cooling tower is usually designed with multiple cells, each operating with the same water to air flow rate, there would not only be a savings in fan energy but also an altered evaporation rate for the given heat load. Although there are noticeable differences between the results for the bypass case and those of Fig. 4-2 during the colder months, these differences are found to be small when averaged over the year and can be neglected for our purposes.

Before applying the results of Fig. 4-2, an important note of caution is that estimates of the annual water consumed in wet cooling should be based on an average of the month-to-month figures. Thus the heat transferred per unit weight of water evaporated should be calculated for each month of the year, using the

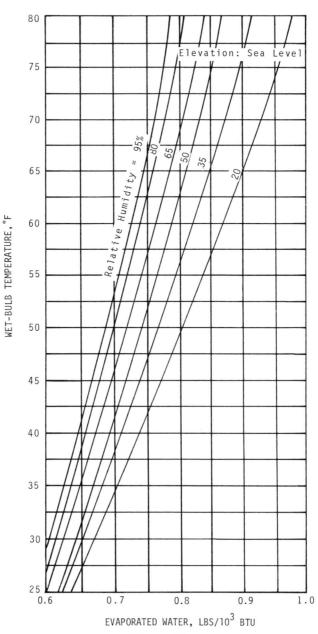

Figure 4-2
Pounds of water evaporated per 1000 Btu of heat transferred in a
wet cooling tower.

long-term, average ambient conditions appropriate to the given month, and then an average taken for all 12 months. Basing the annual evaporation rate on the average annual ambient conditions, on the other hand, could lead to considerable errors in the results.

Table 4.1 summarizes the annual averages of heat transferred per pound of water evaporated for the major coal bearing regions of the United States (see Table 2.2). The southern Rocky Mountain region also encompasses the major oil shale areas. The data were averaged in each region for the centers of the coal or shale areas located in the states indicated. The results show what the curves of Fig. 4-2 anticipate: that in the colder and drier areas the greatest amount of heat is transferred per unit weight of water evaporated, while in the warmer areas the least amount is transferred. Of course, an even more important point is that the difference between the highest and lowest values is only about 8 percent. An estimating figure of 1400 Btu/lb evaporated would be quite satisfactory for the areas with the greatest coal and shale resources.

In a mechanical draft tower, the ratio of water to air flow rates is nearly constant. On the other hand, in a natural draft tower, air flow rates average about 140 percent more in winter than in summer.[2] A natural draft tower designed for summer conditions therefore has a lower water-to-air flow ratio in the winter. For a given cooling load this means the evaporative loss is less in the winter, a direct result of the increase in sensible heat transfer to the large volume of air flow. Leung,[2] however, observed that the annual evaporative water loss in a natural draft tower is only about 3 percent lower than in the mechanical draft tower. This relatively small difference also agrees with the small differences that result if bypassing were practiced.

Table 4.1
Annual averages of heat transferred per unit weight of water evaporated in wet cooling towers for the major coal and shale regions of the United States.

Coal or Shale Region	States	Btu Transferred/lb Evap.
Northern Great Plains	Montana, North Dakota, Wyoming	1,430
Appalachian Basin	West Virginia, western Pennsylvania, eastern Ohio and Kentucky	1,390
Illinois Basin	Illinois	1,390
Southern Rocky Mountains	Colorado, New Mexico, Utah	1,320

Makeup water is required to replace the water that is evaporated. Also, there is always a finite concentration of dissolved impurities in the makeup and these impurities become further concentrated in the circulating cooling water because of the evaporation. One method of preventing the buildup in concentration of impurities to the point where the water becomes saturated and the impurities settle out is to continuously remove a certain fraction of the circulating cooling water and replace it with a cleaner makeup water. Such "blowing down" of the circulating cooling water prevents impurity accumulations that would eventually lead to scaling, fouling, or corrosion on the water side of the heat transfer surfaces.

The concentration of a particular constituent of the circulating cooling water C_b depends on the concentration of the constituent C_m and the flow rate Q_m of the makeup, the evaporation rate Q_e, the blowdown rate Q_b, and the drift rate Q_d. Mass conservation of the constituents entering and leaving the cooling tower (see Fig. 4-1) gives

$$Q_m C_m = Q_b C_b + Q_d C_b. \qquad (4.1)$$

The water balance around the tower is

$$Q_m = Q_e + Q_b + Q_d. \qquad (4.2)$$

There is an approximate direct relationship between the "number of cycles" to which the circulating cooling water can be concentrated, as measured by the ratio of the circulating water concentration to the makeup water concentration, and the ratio of the makeup to blowdown rates. The number of cycles of concentration N is obtained by combining Eqs. (4.1) and (4.2) to give

$$N \equiv \frac{C_b}{C_m} = \frac{Q_m}{Q_b + Q_d} = \frac{Q_m/Q_e}{Q_m/Q_e - 1} . \qquad (4.3)$$

In a well designed cooling tower, the drift rate Q_d is generally very small, around 0.005 percent of the circulating cooling flow rate or about 0.25 percent of the evaporation rate,[3] and can thus be neglected to give

$$N \approx \frac{Q_m}{Q_b} . \qquad (4.4)$$

By applying Eq. (4.3) when $N = 2$, 5, and 10, the ratio Q_m/Q_e is found to be 2, 5/4, and 10/9, respectively. Therefore, a cooling tower should be operated at the highest cycle of concentration possible to minimize the ratio of the makeup to evaporation rate and also, as shown by Eq. (4.4), to reduce the blowdown rate. The limit to the cycles of concentration is set by such problems as scaling and fouling. Certain constituents in the water may be added or removed from the cooling tower by ways other than through the makeup or blowdown. For example, sulfates may be added to the tower by scrubbing sulfur dioxide from the air or by the addition of sulfuric acid to drive out dissolved carbon dioxide; hardness may be removed by chemical treatment in a sidestream. The concentration of a particular constituent in the tower can be determined by mass conservation. If the mass flow rates of a given constituent in and out of the tower, exclusive of makeup and blowdown, are M_i and M_o, respectively, then the concentration of the constituent in the tower is given by

$$C_b = \frac{Q_m C_m + M_i - M_o}{Q_b} .$$ (4.5)

The term "cycles of concentration" for a tower utilizing sidestream treatment refers to soluble materials only, since insoluble materials are removed in the sidestream treatment. The blowdown flow rate is determined by dividing the makeup flow rate by the cycles of concentration.

4.2 Dry and Wet/Dry Cooling Systems

One means to eliminate the makeup water requirements associated with evaporative cooling towers is to use large finned-tube heat exchangers which are air cooled. This is termed dry cooling or air cooling. Air can be drawn past these heat exchangers by a forced or natural draft. For process purposes in synthetic fuel plants the many points of cooling load are spread out, and so the natural draft tower would not be appropriate. Therefore, we shall consider only forced air cooling units.

The method of transferring heat from the hot stream to the air may be a direct or indirect one. In the direct system the air cooled heat exchangers reject heat directly to the atmosphere from the condensing steam or hot process stream. There are two indirect systems. In one, a circulating cooling water loop is in contact with the hot stream and then cooled. This system may be appropriate when all the cooling is in one place. In synthetic fuel plants this is not an advantage and the penalties associated with having two heat transfer surfaces in series, and hence a larger overall temperature difference, rule out its use. In the other indirect system,[4] the cool circulating water is sprayed into the condenser where it mixes with and absorbs heat from steam as condensation takes place.

Part of the hot condensate then circulates to the air cooled heat exchanger, and
the remaining part returns to the boiler. Clearly, the circulating water in this
system must be a high quality condensate. Thus the system cannot be applied to
cooling process streams.

A dry cooling unit operates entirely on sensible heat transfer to the air. The
heat exchanger surface area to reject a given amount of heat is related to the
difference between the ambient dry-bulb temperature and the temperature to
which the process stream is cooled. In warm weather, wet evaporative cooling
towers can cool to a lower temperature than dry coolers can. In cold weather,
however, an air cooler may give a much lower process stream temperature if this
is desirable, because circulating cooling water must not be allowed to freeze. The
control of the process stream temperature is necessarily more complicated with a
dry cooler, since the dry-bulb temperature has a considerably wider daily variation than does the wet-bulb temperature.

In general, for a given heat removal, dry coolers will have a higher cost than
wet coolers because a larger heat exchanger surface is required. Moreover, this
cost difference increases as the temperature level at which the heat is removed is
lowered. It must also be remembered that capital cost involves not only the cost
of heat exchangers and cooling tower assemblies but also those pieces of plant
equipment which depend on the temperature of cooling. For example, the horsepower of a compressor is directly proportional to the absolute temperature of the
entering gas. If the gas is cooled down to a lower temperature between stages of
compression (a necessary procedure in the production of oxygen), then less
horsepower and hence a smaller capital investment would be required.

Interesting methods have been tested to reduce the capital cost of dry coolers
by alternative concepts of heat exchanger design and operation, although they
have not yet been commercially developed. Included among these alternatives is
the fluidized bed system,[5-7] in which the finned cooling tubes are embedded in a
shallow bed of particles that become fluidized when air is drawn through the bed.
The agitation of the fluidized particles increases the heat transfer coefficient of
the cooling tubes, which permits a smaller heat exchanger to transfer a given
amount of heat. However, a penalty is paid in the form of the increased power
required to compensate for the pressure drop of the air through the bed. This
system can also be operated in a wet mode by evaporating dirty water sprayed on
the bed. With no chemical treatment of water and no blowdown, the operating
cost may be reduced and the water usage lessened.

Another interesting but still undeveloped system is the rotating disc cooling
system.[8] Here the cooling water is circulated through an oil-covered pool through

which a number of half immersed large discs mounted on a central shaft slowly rotate. The heated discs rotate to the air side where they are cooled by a stream of air blown through the channels formed by the parallel mounted discs. The layer of oil on the water surface coats the discs as they leave the water and prevents evaporation by eliminating a direct air-water interface. The discs are far less expensive to fabricate than conventional heat transfer surfaces. By removing the oil, this system can also be operated in a evaporative mode.

A combined wet/dry cooling system may be a cheaper alternative to a conventional dry cooler and have a lower water consumption rate than a wet tower.[9] An example of such a system applied to a steam condenser is shown in Fig. 4-3. The configuration illustrated here is not unique and other alternative arrangements are possible. Under peak summer load conditions the system would be designed to divide the cooling load into a predetermined proportion between a dry and wet condenser.

For purposes of this discussion let us say that we wish to hold the condenser temperature T_c in the range of 115° to 135°F. In that case the design point for the condenser temperature is 135°F with all the fans in the dry condenser and all the units around the wet tower operational. Also, let us say that without any bypass of water to the wet tower, the inlet cold water temperature to the condenser, T_{wc}, is 100°F and the exiting hot water temperature, T_{wh}, is 125°F. As the ambient temperature drops during the year, both the dry cooler and the wet tower are allowed to operate at full power. The condenser temperature eventually drops to

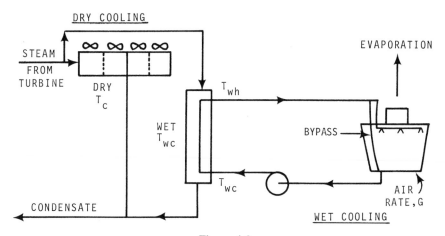

Figure 4-3
A combined wet/dry system for cooling a steam condenser.

115°F when the ambient temperature is cold enough. If the ambient temperature continues to drop, the units in the wet tower are gradually shut off (the dry section is still operating at full power) so that the condenser temperature remains at 115°F. This shifts the condensing load from the wet tower to the dry condenser and reduces the amount of water evaporated in the tower. As the ambient temperature continues to drop during the year, more wet units are shut off until eventually the wet tower is completely shut down. If the ambient temperature continues to drop further, fan power in the dry condenser is reduced to maintain the condenser temperature at 115°F. The operating procedure described here for the wet and dry cooling units, with respect to changing ambient conditions, is also appropriate to each of these units when installed by themselves. Water consumption with a wet/dry tower depends on how the wet tower is operated with respect to the air and water flow rates for various loads and ambient conditions, and thus no specific rule can be given.

Laranoff and Foster[9] have evaluated combined wet/dry cooling systems with different "tower size ratios." These systems are designed to reduce the makeup water in the cooling of steam-electric turbine condensers. The tower size ratio, as they define it, represents the ratio of the heat load split between the wet tower and the dry tower at the design ambient temperature. For example, a 75 percent size wet tower dissipates 75 percent of the total heat load, and a 25 percent size dry tower dissipates the remaining 25 percent of the heat. In addition, without optimizing costs, they examined four operating modes for the wet tower. In one limit they required an infinite control of the water and air flow rates leading to a minimum consumption of makeup water. In the other limit, neither control of the air nor of the water flow was provided. When the ambient temperature dropped to a specified value (70°F), the wet tower was assumed to be removed from service. This operating scheme required the greatest amount of water and fan power.

For different tower size ratios and different operating modes, Laranoff and Foster calculated the fraction of makeup water that would be used annually in the combined system and compared that to the fraction for an all wet tower. These calculations were made for 18 selected cities covering all parts of the United States. Their results show that for a given configuration there is not a significant difference in the fractional makeup water requirement among the different coal and shale regions. Rather, the largest differences arise from the operating mode chosen. In Fig. 4-4 Laranoff and Foster's results are plotted for the annual fractional makeup water, averaged between the minimum and maximum values imposed by the choice of control, then further averaged with respect to the

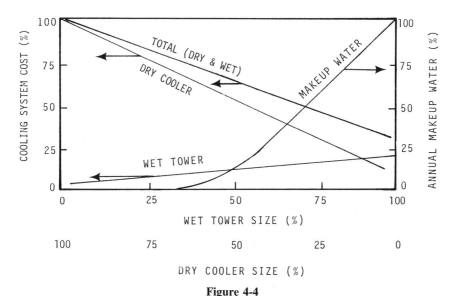

Figure 4-4
Relative costs and average annual makeup water requirements for combination wet/dry
cooling systems used on steam-electric turbine condensers in coal and shale regions.

slightly differing results obtained at the different selected sites in the coal and
shale regions. Also plotted in Fig. 4-4 are curves taken from Ref. 9 showing the
relative costs of the dry and wet tower components, their sum representing the
total system cost relative to the all-dry cooling system. These results may be
considered representative for steam condenser applications in synthetic fuel
plants since they are fractional and not absolute values. The curves show that
water consumption can be reduced to about 25 percent of that required for all wet
cooling by making a capital investment less than 60 percent of that required for
all-dry cooling.

4.3 Cooling Ponds and the Cost of Water

In once-through cooling, water is withdrawn from a large natural water body—an
ocean, a river, or a lake—is circulated through the heat exchanger, and is subsequently discharged to the receiving water where the heat is dissipated. This
method of cooling is presently the most important means of cooling steam-
electric power plants. Thermally, it is the most efficient method of disposing of
waste heat and, in those areas where sufficient water is available, the most
economic. Because of the vastly greater surface area of the water in comparison

with a wet evaporative tower, there is a greater heat loss to the atmosphere by convection and radiation (processes that do not consume water) and less heat loss by evaporation. Because of environmental restrictions on thermal pollution, however, this method of cooling is expected to be increasingly restricted, even near ocean sites. Since we are concerned with the interior coal bearing regions of the United States, we shall not consider once-through cooling further.

An essentially identical, but environmentally acceptable, method is to use a cooling pond. A cooling pond is an impounded body of water built on or near the plant site for the purpose of serving as a heat sink. Warm water is discharged from the condensers to the cooling pond where it is cooled and recirculated back through the condensers. The condenser temperature is related to the pond size. Too small a pond can result in too high a pond temperature and therefore too high a condenser temperature. The surface area for cooling ponds used for cooling steam-electric power turbine condensers lies in the range of one to five acres per megawatt of electricity generated, with 1.5 to 2 acres/MW often used for exemplary calculations; the actual value is very dependent on site. For electric power generation, a 33 percent efficiency translates 2 acres/MW generated into about 1 acre/MW of heat dissipated, since about two units of heat are dissipated for each unit of heat converted into electricity. Land costs are a major factor in the cost of a cooling pond.

A cooling pond has a number of advantages over a closed-cycle wet cooling tower. It is cheap to install in those areas where land is relatively cheap and it requires little maintenance.[10] A cooling pond also has a high thermal inertia that enables the inlet temperature to be maintained at a reasonably constant value, independent of short-term changes in ambient conditions. Although cooling ponds do not have the problem of drift, they do generate fog over and near the pond.

The advantages of a cooling pond over a wet cooling tower must be balanced against the fact that a cooling pond has a higher evaporation rate. Figure 4-5 shows the evaporative water loss, normalized with respect to the waste heat dissipated, as a function of the wet-bulb temperature.[11] At a wet-bulb temperature of 70°F, characteristic of summer conditions in the southern part of the Appalachian Basin, the evaporative water loss from a cooling pond at a loading of 2 acres/MW of electricity generated is 60 percent higher than from a mechanical draft wet cooling tower. For conditions of the Northern Great Plains, the evaporation rate from a cooling pond in the summer is about 50 percent higher than from a cooling tower.

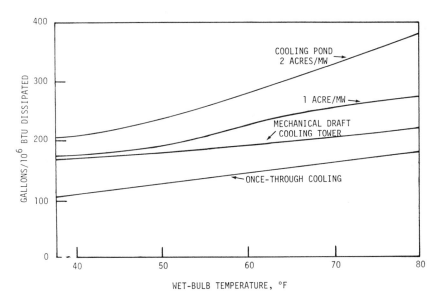

Figure 4-5
Water evaporated as a function of wet-bulb temperature for different cooling methods.

The desirability of a pond in place of a wet cooling tower is dependent on the trade-off between the cost of water and the cost of money. The cost of water is made up of a number of components. In most western states water must be bought from its source. A modest cost for water rights is in the range of $10 to $20/acre-ft, which is equivalent to from $0.03 to $0.06/10³ gal. In the eastern states there is generally no direct cost, provided there is accessibility to the water, which in itself may involve a cost. The cost of moving water to a site is about 1 to 2 cents per thousand gallons per mile. Again, this would be higher in the West than in the East, since the distances to the water sources tend to be greater. For most cases, however, the distances are not greatly different. Finally, circulating water must be treated, and for wet evaporative towers the minimum cost is $0.20/10³ gal evaporated. For comparative purposes, a raw water cost of $0.40/10³ gal has been recommended.[12] This might be the cost of water rights plus transportation from 30 miles, or water rights plus uphill pumping from a lake or pumping from underground. When treatment is added, cooling water can often cost $0.60/10³ gal evaporated. This is probably the low end of the cost. As the distance from the water source to the plant increases to 100 miles or more, the cost of water rights increases, or the water becomes more contaminated (necessitating more expensive treatment), then the cost of water evaporated for cooling can approach $2/10³ gal evaporated. For purposes of discussion in this text, we shall consider cooling water to cost in the range of $0.60 to $2/10³ gal evaporated.

With the above figures, we may estimate whether the saving in water costs justifies the use of a wet cooling tower. Although this criterion can be quantified, it is not the only one; for example, social and political criteria could be more important. The difference in capital cost between a mechanical draft wet tower and a cooling pond is about $1 to $3/kw electricity generated,[10] corresponding to about 0.25 to 0.75 cents per million Btu of heat dissipated, for an amortization of 15 percent per year. If we use 50 percent more water in a cooling pond than in a cooling tower, and if one pound of water is evaporated per 1400 Btu transferred in a wet tower (see Table 4.1), then based on the results cited above 0.5 pounds of water are saved for every 1400 Btu dissipated by means of a cooling tower. For the cooling tower to pay for itself, the cost of cooling water saved must be greater than $0.06 to $0.18/10³ gal evaporated. If cooling water costs are taken to be in the range of $0.60 to $2/10³ gal evaporated, it follows that cooling towers are more economically advantageous than cooling ponds in the coal and oil shale regions of the country. Of course, if the water supply is not reliable, large reservoirs may have to be built to store water for the operation of the cooling tower. Under these conditions it might then be cheaper to use the reservoirs as cooling ponds and dispense with the cooling towers.

A variant on cooling ponds is "phased cooling."[13] This method takes advantages of the differences between day and night ambient conditions. It employs relatively small hot and cold water storage ponds and a relatively large but shallow cooling surface. In the first phase, during the hottest hours of the day, cooling water is taken from the cold pond and then after circulating through the heat exchanger is discharged into the hot pond. The second phase is used at night to obtain maximum cooling by convection and radiation, thus minimizing evaporation. In this phase the cooling water is circulated between the shallow cooling surface and the condenser. At the same time, water flows from the hot pond over the cooling surface and into the cold pond. The water saved is at least as much as in a simple pond, and it may be possible by this technique to effect a substantial reduction in the land area requirements. However, the cooling surface pond may have to be lined, and this could significantly add to the cost.

4.4 Cooling Process Streams and Gas Purifiers

In this section and in the remainder of the book our presentation of cooling systems will be limited to mechanical draft wet cooling towers, mechanical draft finned tube dry heat exchangers, and the use of both units in a combined system. Our reasons for these choices are not only the limitations on the use of natural draft towers and cooling ponds previously noted but, more importantly, the fact that by themselves the forced circulation wet tower and dry cooler are standard pieces of equipment in the chemical processing industry. Their cost is known and their operating conditions are well defined. With information of this quality for these systems, achievable technology can be presented. Some other methods described, however, could also lead to a decreased water consumption but their costs cannot now be precisely given. Moreover, the proper use of the forced circulation wet tower and dry cooler can in many instances lead to a low consumption of water, at a moderate cost and under a wide variety of conditions.

In the introductory discussions some of the major points of cooling in a synthetic fuel plant were noted. In the next two chapters the description of the various technologies will include the type of thermal loads each carry. For our purposes here it is sufficient to recognize that there are at most four principal types of cooling loads in any synthetic fuel plant. These loads are process streams, gas purification, turbine condensers, and gas compressors. Our discussion on the criteria for defining the appropriate methods for cooling these loads begins with the consideration of process streams.

Below the range 270° to 300°F, useful heat cannot be extracted from a process stream, and instead further cooling is needed to dissipate the heat to the atmosphere. A simple calculation outlined below shows that the most economic procedure is probably to cool to about 130° to 140°F using an air cooler and to cool below that temperature using a wet system. Again, there is an economic trade-off dependent on the process conditions between the cost of water and the cost of money.

The cost of forced circulation dry cooling includes the cost of capital, which is principally the charge for the heat transfer surface, and the cost of energy, which is principally the charge for running the cooling fans. Because a wet cooling tower also evaporates water, there is another cost in this system related to the cost of water. The capital costs in the wet system include the charge for the heat transfer surface and the charge for the tower, which depends on the amount of water circulated through it. Correspondingly, the energy cost is made up of two

components: the charge for operating the tower fan and the charge for pumping
the circulating water.

The above principles may be expressed in relatively simple cost equations for
dry and wet coolers. For a dry cooler the total cost per unit of heat transferred, C_d
(in $/10^6$ Btu, for example), is

$$C_d = k_c \left(\frac{A}{q}\right) + k_p \left(\frac{P_d A}{q}\right), \tag{4.6}$$

where the use of consistent units is always assumed. Here, q is the rate of heat
transfer, A is the heat transfer surface area, and P_d is the power per unit area of
heat transfer surface to run the cooling fans. The cost coefficient k_c is the
amortized cost per unit area of heat transfer surface and k_p is the cost of power.
The heat transfer rate can be expressed in terms of a heat transfer coefficient
through the standard relation

$$q = U_d A \Delta T_d, \tag{4.7}$$

where U_d is the dry cooler heat transfer coefficient and ΔT_d is the log mean
temperature difference across the cooler. With Eq. (4.7) the cost equation (4.6)
can be rewritten in the more useful form

$$C_d = \frac{1}{U_d \Delta T_d} (k_c + k_p P_d). \tag{4.8}$$

A similar equation to Eq. (4.6) can be written for the wet system, which also
includes the added charges proportional to the circulating water rate and the
charge for water that is evaporated. The total cost per unit of heat transferred,
C_w, can be written

$$C_w = \left[k_{c1} \left(\frac{A}{q}\right) + k_{c2} \left(\frac{M_{circ}}{q}\right)\right] + k_p P_w \left(\frac{M_{circ}}{q}\right) + k_w \left(\frac{M_{evap}}{q}\right). \tag{4.9}$$

Here, M_{circ} is the circulating water rate, M_{evap} is the evaporation rate in the
tower, and P_w is the power per unit circulating flow rate to operate the tower fans
and circulating pumps. The cost coefficient k_{c1} is the amortized cost per unit area
of heat transfer surface, k_{c2} is the amortized tower cost per unit circulating flow
rate, and k_w is the cost of water, treated here as a parameter. The generalized heat
transfer rate relation is given by Eq. (4.7). For conditions appropriate to wet heat
transfer, the parameters are denoted by the subscript w. The circulating water

rate can be expressed in terms of the cooling range across the heat exchanger, here denoted by Δt_w, through the relation

$$q = M_{circ} \, c \, \Delta t_w, \qquad (4.10)$$

where c is the specific heat for water (1 Btu/lb-°F). In Section 4.1 we showed that a suitable estimating value for the evaporative water rate was

$$\frac{M_{evap}}{q} = \frac{1 \text{ lb evaporated}}{1400 \text{ Btu transferred}} . \qquad (4.11)$$

Using Eqs. (4.10) and (4.11), the cost equation for a wet tower becomes

$$C_w = \frac{k_{c1}}{U_w \Delta T_w} + \frac{1}{c \Delta t_w} \, (k_{c2} + k_p P_w) + \frac{k_w}{1400 \text{ Btu/lb}} , \qquad (4.12)$$

with consistent units assumed.

The calculation of the dry and wet log mean temperature differences is straightforward,[14] once the air temperature rise across the dry cooler and the water temperature rise across the wet cooler are known. Here and throughout the book, for estimating purposes, a water temperature rise of $\Delta t_w = 25°F$ is generally assumed. In actual designs this cooling range may be controlled by a limit on maximum turbine back pressure or by a tendency to scale in the cooling water circuit. An empirical formula for estimating the air temperature rise is given by[15]

$$\Delta t_d = 0.005 \; U_d \left(\frac{T_1 + T_2}{2} - t_1 \right) , \qquad (4.13)$$

where t_1 is the inlet air temperature, and T_1 and T_2 are the process stream inlet and outlet temperatures. In actual designs this temperature rise would be optimized.

Both Eqs. (4.11) and (4.12) are quite general and suitable for estimating purposes. The heat transfer coefficients and power requirements appearing in these equations are reasonably well defined. In Table 4.2 representative heat transfer coefficients are shown for streams in synthetic fuel plants. Representative power requirements for the fans to move the air in dry and wet towers and for the pumps to circulate the water in a wet tower are given in Table 4.3.[15]

The only quantities that may be expected to vary with future conditions are the cost coefficients. Table 4.4 shows present representative values; they could be revised, however, without altering the validity of the general relations presented. The capital cost coefficients are for installed, bare equipment and are suitable for comparing wet and dry cooling. They should not be used for estimating absolute

Table 4.2
Representative heat transfer coefficients for dry and wet cooling.

Stream	U (Btu/hr-ft^2-°F)* Dry	Wet
Gas compressor interstage coolers†	10–50	12–70
Cooling low pressure gas stream	40	50
Cooling high pressure gas stream	70	100
Condensing steam in gas purifier	75	110
Condensing steam from turbine drives	120	170
Cooling water for gas scrubbing	130	400

*Bare tube area for dry cooler finned tubes.
†Ranges correspond to from 10 to greater than 500 psig.

Table 4.3
Representative power requirements for dry and wet towers.

Units	P_d (hp/ft^2)	P_w (hp/gpm)
Fans		
$U_d \geqslant 100$	0.020	—
$50 \leqslant U_d < 100$	0.0175	—
$U_d < 50$	0.015	—
Wet	—	0.012
Pumps		
Wet	—	0.033

Table 4.4
Cost coefficients in relations for cost per unit of heat transferred in dry and wet cooling towers.

Cost Coefficient	Dry	Wet
k_p (¢/kwh)	2	2
k_{c1} ($/ft^2/yr)	3	1–1.80*
k_{c2} ($/gpm/yr)	—	1

*Less than 300 to greater than 600 psig.

plant costs without including a suitable factor to account for piping, instrumentation, engineering, and other costs.

By way of example, to illustrate the trade-off with the price of water, we consider the cost of cooling gas streams, one at high pressure and one at low pressure, for removing a unit heat load at a given rate from 280°F to a specified exit temperature. A summer design condition is chosen with the ambient air temperature at 90°F and with cooling water available at 80°F. The results of the calculations using Eqs. (4.8) and (4.12) and the data in Tables 4.2 to 4.4 are plotted in Figs. 4-6 and 4-7. It can be seen that the choice of wet or dry cooling depends on the process conditions, as well as the cost of water. Therefore, the calculations generally must be repeated for each individual situation. Such detail is beyond the scope of this book, but our results do indicate a sufficiently accurate estimating rule: all process streams should be cooled to about 130° to 140°F using dry cooling and below this using wet cooling.

Removal of the acid gases, carbon dioxide, and hydrogen sulfide is an impor-

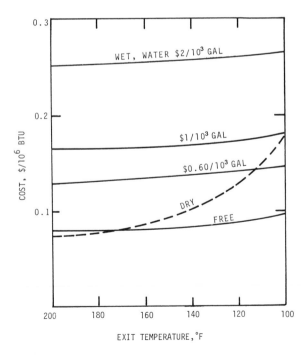

Figure 4-6
Cost of cooling a high pressure gas stream.

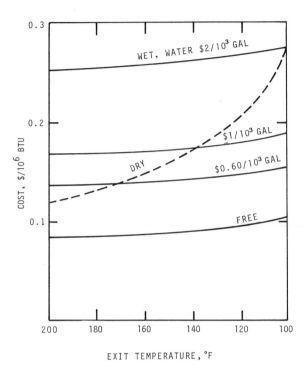

Figure 4-7
Cost of cooling a low pressure gas stream.

tant energy-consuming part of synthetic fuel plants, and the regenerator condenser for this purpose is one of the major cooling loads (see Section 3.6).
Whether dry or wet cooling is used will depend on the gas purification process.
Two types of gas purifiers are considered here, one that uses hot potassium
carbonate, of which the Benfield process[16] is representative, and one that uses an
organic solvent, of which the Selexol process[16] may be considered representative. In both systems, the condenser in the stripping column (see Fig. 3-4) must
remove an amount of heat about equal to the heating load previously given in
Table 3.5. The feed stream to the condenser is a mixture of water vapor and the
noncondensable acid gases, most of which is carbon dioxide. These vapors enter
the condenser at about 230°F and, as the mixture is cooled, condensation proceeds.

For a hot potassium carbonate system, cooling to 140°F will condense enough
water to keep the circulating absorbent in balance. The heat transfer coefficients

for both dry and wet cooling are given in Table 4.2. From Eqs. (4.8), (4.12), and the data in Tables 4.3 and 4.4, the dry and wet cooling costs may be calculated as a function of the price of water. It is found that air cooling costs $0.12/10^6 Btu and the same cost is found for wet cooling if water costs $0.46/10^3 gal. Having estimated that the cost of water (including the cost of supply, treatment, and blowdown disposal) will have a minimum price of $0.60/10^3 gal, then dry condensing is clearly the most appropriate for this system.

For an organic solvent type process, cooling to about 100°F is probably necessary to keep the water system in balance (see Section 3.6). By air cooling to 130°F and then wet cooling to 100°F, 90 percent of the total cooling load is taken by the air cooler and 10 percent by the wet cooler.

4.5 Cooling Turbine Condensers and Gas Compressors

All synthetic fuel plants consume a significant amount of energy in mechanical drives. Large drivers are needed for such uses as gas compression, circulating cooling water, circulating acid gas absorbing liquid, and pumping coal slurries into high pressure reactors. Most plants will generally be self-sufficient so that electricity will be generated on site. It makes no difference in this discussion whether a turbine driver is used to generate electricity and an electric motor then drives a gas compressor, or whether the turbine driver is attached directly to the gas compressor. In either case, the energy requirement and cooling load are not much different. However, the capital cost is increased by the use of electric motors. In practice, direct turbine drivers are used on most machinery and electrical generation is kept to a minimum.

The working fluid for turbine drivers can be a fuel gas or steam. If steam is used it may be condensed as it leaves the turbine and the condensate returned to the boiler. Alternatively, high pressure, superheated steam may be fed to the turbines with low pressure, saturated steam removed from them and condensed in heat exchangers in the process plant wherever low temperature heat is required. All of these procedures have been and will be used. In synthetic fuel processes, a good deal of heat is emitted which can best be recovered by producing steam in heat exchangers. Typically, all the low level heat needed in the process will be produced in other parts of the process, and steam driven turbines will be designed with condensers.

Gas turbines will be used if the fuel used to drive the plant is a gas. The major reason for making and burning fuel gas on site is that removal of sulfur in the form of hydrogen sulfide from a fuel gas is easier than removal of sulfur in the form of sulfur dioxide from the stack of a coal or char burning furnace. Producing and burning a fuel gas is probably more expensive in capital and energy than burning coal or char to raise steam in a boiler, even when flue gas desulfurization is required. Current designs include both possibilities, but most designs rely on condensing steam turbine drivers; this type of driver is assumed in all the examples given in this book.

Steam turbine condensers may be wet cooled, dry cooled, or have a wet and dry cooler in parallel as shown in Fig. 4-3. The costs that govern the choice of cooling system include all the factors given so far, plus the efficiency characteristics of the turbine. For a given turbine design the power available at the shaft (in kw, for example) requires a certain steam heat input rate (in Btu/hr). The ratio Btu/kwh is commonly referred to as the "heat rate" of the turbine and is related to the turbine efficiency, η_{turb}, by the expression

$$\eta_{turb} = \frac{\text{Ideal Heat Rate}}{\text{Actual Heat Rate}}, \qquad (4.14a)$$

where

$$\text{Actual Heat Rate} = \text{Ideal Heat Rate} + \text{Dissipated Heat Rate}, \quad (4.14b)$$

and

$$\text{Ideal Heat Rate} = 3,413 \text{ Btu/kwh}. \qquad (4.14c)$$

An actual heat rate of 10,000 Btu/kwh therefore corresponds to an efficiency of 34 percent.

The turbine efficiency depends on its design, the condition of the steam produced by the boiler, and the condenser temperature. Large turbines for coal fired electric generating are designed for superheated, high pressure steam at temperatures of about 1,000°F. The steam pressures are at least 1,500 psig and go up to 4,000 psig, which is above the critical point. For a fixed design and steam conditions, the efficiency depends on the condenser temperature, which at 120°F (corresponding to 3.5 inches of mercury absolute pressure) is about 40 to 45 percent. The efficiency increases slightly as the temperature falls below 120°F and decreases as the condenser temperature rises. A plot of the dependence of the heat rate on condenser temperature is termed the "turbine characteristic." For the type of turbine described, the turbine characteristic is shown in Fig. 4-8 by the curve labelled "generator type." Note that the heat rate is inversely related to the efficiency.

Industrial type turbines are designed for lower steam pressures and tempera-

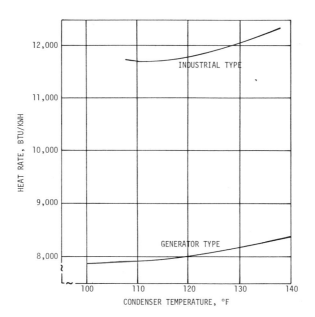

Figure 4-8
Steam turbine characteristics.

tures. The characteristic shown in Fig. 4-8 is for steam in the range between 700°
to 900°F and 700 to 900 psig. Saturated steam at these pressures has a temperature of 500° to 530°F. The steam might, for example, be generated from the large
amount of heat rejected in the methanation process, for which the heat of reaction
is quite high (see Eq. 3.9). The steam would then be superheated in a boiler. Heat
rejected from this process cannot be converted into steam of any higher pressure.
The penalty for using lower pressure steam is lower efficiency, about 28 percent.
The characteristic curve is quite flat for condenser temperatures between 115°
and 135°F. However, the efficiency falls off (the heat rate increases) below about
115°F and above 135°F. For this reason, a cooling tower in the winter would
probably be operated bypassed when using industrial type turbines, as discussed
in Section 4-2. Note also that since dry cooling will never yield as low a condenser temperature as wet cooling, the added inefficiency is an added cost for dry
cooling.

The cost of cooling steam turbine condensers cannot be expressed in a single
formula. Each part of the cost must be determined separately, operating costs
determined separately for each month, and annual costs taken as the sum of the
components. We next consider the parallel wet/dry cooling system illustrated in

Fig. 4-3 and outline a procedure for determining the total annual cooling cost as it depends on the cost of water.

First, we must assume that a fraction of the cooling load is carried by the dry condenser. To find the most economical fraction, the calculation must be repeated for different assumed fractions. For example, suppose we take as a basis 1 kw-yr of delivered shaft energy, with the condenser 50 percent dry cooled and 50 percent wet cooled. A typical set of climatic information for Farmington, New Mexico, is shown in Table 4.5. Also shown in Table 4.5 are the "design" temperatures for this example. The design temperatures are selected on the assumption that they will not be exceeded for more than ten hours in an average year. The turbine characteristic is also needed, and we suppose this to be the industrial type shown in Fig. 4-8. The condenser is taken to be cooled to 135°F at design conditions, consistent with the example in Section 4.2.

From Fig. 4-8, the heat rate for a condenser temperature of 135°F is 12,200 Btu/kwh. The total cooling load is given by Eq. (4.14b) to be about 8,800 Btu/kwh. The dry cooling load is half of this or 4,400 Btu/kwh. Using Eqs. (4.7) and (4.13), the size of the dry condenser is found to equal 1.5 ft² in our example. Knowing this area, the capital cost of the dry condenser can be determined. To size the wet condenser, the inlet and outlet temperatures must be known for the

Table 4.5
Monthly average temperatures at Farmington,
New Mexico.

	Dry-Bulb Temperature (°F)	Wet-Bulb Temperature (°F)
January	26	23
February	33	28
March	42	33
April	49	37
May	60	45
June	70	51
July	76	58
August	73	57
September	64	49
October	51	41
November	39	32
December	27	24
Design	98	65

circulating cooling water. Following the illustrative example in Section 4.2, the inlet cold water temperature is taken to be 100°F and the exiting hot water temperature to be 125°F. The area, and hence the capital cost of the wet condenser, can be found from an equation similar to Eq. (4.7). The rate of flow of circulating cooling water can be found from Eq. (4.10) and hence the capital cost of the cooling tower can be determined.

The capital investment is now known, as is the size of the dry and wet condensers. We also need the operating cost, however, and this depends on the cooling tower performance when the ambient conditions are other than design conditions. To find the operating cost, month by month calculations are required. The procedure begins with the dry condenser. The average dry-bulb temperature for the hottest month of the year is used. The condenser is assumed to be cooled to a specified temperature which must then be verified by calculation. The temperature of 115°F is initially chosen because it corresponds to the minimum heat rate and hence cooling load (see Fig. 4-8). From the turbine characteristic, the total cooling load of 8,300 Btu/kwh is found. Knowing the area of the dry condenser, its cooling load rate can be found from Eq. (4.7) for the temperatures assumed, which for our example is 4,600 Btu/hr. The cooling load not carried by the dry condenser must be carried by the wet condenser.

The calculation of the load that the wet condenser can carry cannot be laid out in an analytic form as simple as that for the dry condenser. In addition to the wet condenser area the load also depends on the performance characteristics of the wet tower. The type of tower chosen does not matter, since tower characteristics are needed only to determine off-design performance and not construction criteria. It does not matter, for example, whether the tower is a countercurrent tower in which the air passes up and the water down, or a cross-flow tower in which the water passes down and the air moves horizontally. Performance curves and computer programs are available for determining cooling tower performance (see Refs. 17 and 18 for examples), and any consistent set may be used for the procedure outlined here.

The wet calculation begins by recognizing that in practice the rate of circulation of cooling water through the condenser is not varied during the year so long as some wet cooling is required. The main reason is that low flow rates lead to stagnation and fouling by settling of suspended solids and precipitation. The design circulation rate of cooling water is given by Eq. (4.10). For our design wet load of 4,400 Btu/hr and cooling range of 25°F, the rate is 176 lb/hr. Knowing this circulation rate and the off-design cooling load to be handled by the tower (8,300 Btu/hr − 4,600 Btu/hr), the off-design cooling range is determined from Eq. (4.10) to be 21°F. Next, the cooling tower performance curves for this range and the monthly average wet-bulb temperature are used to see what the cold water temperature will be. Finally, the performance of the wet condenser must be checked to see if it will carry the load.

By necessity the calculation is trial and error. If the load cannot be carried then a higher condenser temperature, and hence higher cooling load, must be assumed and the calculation repeated. In fact, in our example, the condenser can be cooled to 115°F. If the load carried is greater than the 8,300 Btu/hr needed then the cooling tower can be bypassed. The next hottest months are then considered. When the monthly average dry-bulb temperature is low enough for the dry condenser to carry the full load, the cooling tower is assumed to be turned off and power to the dry condenser fans reduced.

The monthly cost is the sum of (1) the amortized fraction of the capital investment; (2) the cost of energy for the circulating water pumps (0.033 hp/gpm from Table 4.3), which operate at full power unless the tower is completely turned off when this energy is zero; (3) the cost of energy for the mechanical draft fans in the cooling tower, which from Table 4.3 is 0.102 hp/gpm of water *not* bypassed; (4) the cost of energy for the dry condenser fans, which is 0.020 hp/ft^2 multiplied by the factor given in Fig. 4-9.

A final calculation is needed to estimate the water evaporated. The evaporation rate, Q_e, is given by

$$Q_e = G(H_T - H_B), \qquad (4.15)$$

where G is the air flow rate through the cooling tower and H_B, H_T are the absolute humidities of the air (lb water/lb dry air) at the bottom and top of the tower. The rate of air flow is determined from the tower performance characteristics. The absolute humidity can be read from a psychometric chart or table, given the wet-bulb and the dry-bulb temperatures. The absolute humidity H_B is, therefore, known for each month. At the top of the tower the air is very close to saturation (that is, the wet-bulb equals the dry-bulb temperature). In finding the temperature at the top of the tower, the wet cooling load, q_w, is given entirely to the air and

$$q_w = G(i_T - i_B),$$

where i is the enthalpy of humid air, that is, the enthalpy in Btu of 1 lb of dry air plus the associated H lb of water vapor.

Knowing the ambient dry-bulb temperature and the absolute humidity H_B, the enthalpy i_B can be calculated from standard psychometric formulae or charts.[19]

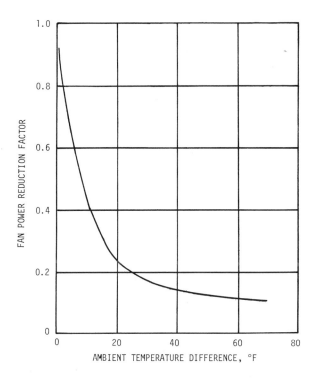

Figure 4-9
Fan power reduction factor as a function of difference
between the actual ambient temperature and the ambient
temperature at which the dry condenser will just carry the full
cooling load with the cooling tower turned off.

The enthalpy of the humid air at the top of the tower i_T can then be calculated
from Eq. (4.15). Using the fact that the air is saturated at the top of the tower and
knowing i_T, the dry-bulb temperature there can be determined. This then enables
the absolute humidity of the air H_T to be deduced, leading to the determination of
the evaporation rate calculated with Eq. (4.15).

Some results of the month by month calculations for Farmington, New
Mexico, are shown in Table 4.6. In Fig. 4-10 the curves summarizing the
relationships between the annual water consumption and cost for different water
charges are shown. Both axes are expressed per unit of kw-yr output shaft
energy. If water is free, a change from all wet to all dry cooling more than
doubles the annual charge, from $5 to $11/kw-yr, and saves just over 5,000
gallons of water. Also shown in Fig. 4-10 is the percentage of the cooling load

Table 4.6
Bi-monthly operating conditions per kw of generated power for a parallel wet/dry
industrial type turbine condenser cooling system at Farmington, New Mexico, with the
condenser 50 percent dry cooled and 50 percent wet cooled at the maximum
design temperature.

	Condenser Temp. (°F)	Dry Cooling Load (Btu/hr)	Power* (10^{-2} kw)		Evaporated Water (lb/hr)
			Dry Fan	Wet Fan	
January	115	8,200	0.51	0	0
March	115	8,200	1.44	0	0
May	115	6,667	2.24	0.11	1.15
July	118	5,168	2.24	0.30	2.50
September	115	6,142	2.24	0.19	1.55
November	115	8,200	1.13	0	0

*Circulation pump power is 0.86×10^{-2} kw when wet tower operates.

carried by the wet condenser at summertime design conditions. From Fig. 4-10, if cooling water costs more than $0.78/$10^3$ gal evaporated, the system should be designed for about 50 percent wet cooling in the summertime, making the annual average water consumption about 10 percent of the all wet system.

Similar curves have been obtained for the colder climates of Beulah, North Dakota, and Casper, Wyoming, and they have all been replotted in Fig. 4-11 to show how the cost of water would control the consumption. The curves of high water consumption of around 5,000 gal/kw-yr represent all wet cooling, and the curves of lower consumption a combined wet/dry system. For the two systems in conjunction, the cost of water at each site is represented by the vertical line. The conclusion is clear: at sites where water is expensive, combined wet and dry cooling systems should be used on steam turbine condensers of the industrial type.

For high efficiency electric generating plants, the use of partial dry cooling of the condenser will only be used at much higher water costs, in the range of $2.00 to $2.50/$10^3$ gal evaporated.[20] The use of combined wet and dry cooling can reduce water consumption to about 25 percent of that for all-wet cooling. Usually the water evaporated for cooling is by far the largest water consumption in an electric generating plant.

In general, in synthetic fuel plants, about half of the energy needed to drive the plant is used to compress gases. In the examples given in the following two chapters, the details of what gases are compressed, why, and how much energy is

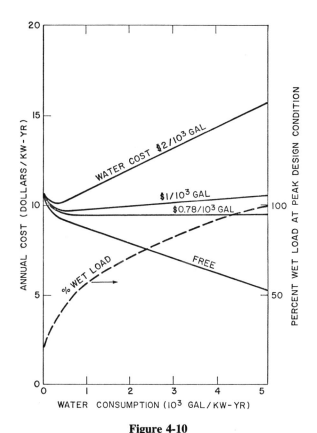

Figure 4-10
Annual cost versus water consumption and percentage of
cooling load carried by wet condenser at summertime design
condition for cooling an industrial type turbine condenser at
Farmington, New Mexico, using a parallel wet/dry system.
The plant operates 7,008 hr/yr (0.8 load factor).

consumed will be described. For the present discussion on cooling we only need
to consider where this energy goes.

When a gas is compressed it heats up. To prevent mechanical damage and
spontaneous explosions caused by the heat, the pressure of a gas is not usually
multiplied more than about 2.3 to 2.6 times without cooling. For this reason gas
compressors are operated in stages. In each stage the gas pressure is multiplied
over the range noted. Between stages the gas is cooled from between 280° and
300°F to between 90° and 120°F. The energy stored in a compressed gas is small

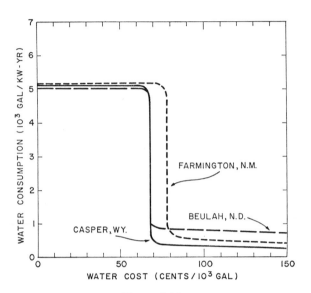

Figure 4-11
The effect of water cost on water consumed for cooling
industrial type turbine condensers.

and more than 95 percent of the energy supplied by the motor or driver to the
compressor is dissipated in the interstage coolers. Interstage coolers may be a wet
system, dry system, or a dry followed by wet system. Each of these have been
used.

The energy needed to compress a given weight of gas through a given pressure
ratio is directly proportional to the absolute temperature of the gas before compression. For example, if gas enters a compressor stage at 160°F, the energy
consumed by that stage will be 1.1 times the energy that would have been
consumed had the gas entered at 100°F. If the interstage cooling does not cool the
gas enough, there will be a penalty cost made up of both the capital to buy larger
compressors and drivers and the energy needed to drive them. Since dry cooling
never cools the gas to as low a temperature as wet cooling, the penalty charge for
higher temperature increases the difference between dry and wet cooling.

Actual calculations involve the number of compression stages and the
molecular weight of the gas, as well as atmospheric conditions and other factors
described in the preceding section. However, the penalty paid for too high an
interstage temperature is such that, as long as cooling water is reasonably priced
and available, wet interstage cooling will be used. Interstage cooling will be the

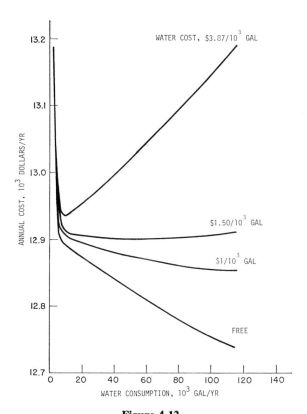

Figure 4-12
Annual cost versus water consumption for air compressor
interstage cooling in New Mexico using a combined wet/dry
system. Basis is 2,000 lb air/hr compressed from
atmospheric pressure to 90 psia.

last place in the plant where dry cooling will be introduced to save cooling water. In Fig. 4-12 the calculations are shown for an air compressor compressing 2,000 lb air/hr from atmospheric pressure to 90 psia. Such compressors are needed to feed air into oxygen production plants and are often the largest compressors in pipeline gas plants. For all-dry interstage cooling the cost is very high, and for this reason some wet cooling will always be used. If water costs $1.50/10³ gal evaporated, it pays to reduce the water consumption to about 10 percent. To do this the gas will, in the summertime, be cooled to 140°F by an air cooler and then to 95°F by a water cooler. In the winter the water cooler will not always be needed.

4.6 A Summary of Cooling Requirements

The quantity of water consumed for cooling will depend on the product quantity, the process, and the location; the cost, availability, and quality of the water; and the economics at the time of design. All that can be done without a full scale plant design is to make reasonable estimates.

The minimum information necessary is the overall plant conversion efficiency. This is defined as

$$\eta = \frac{\text{Heating Value of Product Fuel + Byproducts}}{\text{Heating Value of Raw Fuel}}. \qquad (4.16)$$

Note that all byproducts must be included in the heating value of the materials leaving the plant. In the case of coal gasification, the heating value of the coal fed to the plant is for the coal fed to the boiler plus the coal fed to the gasifier (not the gasifier alone if a boiler is used). If this is all the information available, then in water-rich areas it may be assumed that not more than 50 percent of the unrecovered heat goes to evaporate cooling water at 1,400 Btu dissipated per pound of water evaporated, with 40 percent being a reasonable average. In arid areas or where water is expensive it may be assumed that 25 percent of the unrecovered heat goes to evaporate cooling water and that it is reasonable to reduce this to 15 percent. This gives a very crude estimate for gas plants. For liquefaction and clean coal plants there is so much possible variation in the design that the estimate is subject to considerable uncertainty.

A thermal balance is made around the complete process train to discover (1) the heating value of the coal or shale fed to the process, h_i; (2) the heating value of the product fuel, h_p; (3) the heating value of any byproducts, h_b; (4) the heat recoverable as steam, h_r; and (5) the heat lost to the atmosphere, h_L. As has been shown, it is reasonable for all sites and all qualities of cooling water to assume that process streams will be cooled to between 270° and 300°F with heat recovery, from about 270° to 140°F by air cooling, and below 140°F by water cooling. This allows the division into two portions of the heat lost to the atmosphere through heat exchangers, one to air or dry coolers, h_{Ld}, and one to water or wet coolers, h_{Lw}. In addition, the process design will show how much energy is lost to the atmosphere directly, h_{La}, as in coal drying. The thermal balance will be

$$h_i - h_p - h_b - h_r = h_{Ld} + h_{Lw} + h_{La}. \qquad (4.17)$$

The second step is to discover how much energy is consumed to drive the plant. This energy may be consumed directly as fuel, h_f, as steam for heating, h_s, or in a mechanical form that may be assumed to be as steam for turbine drives, h_m. The total driving energy, $h_f + h_s + h_m$, will be supplied first by using energy recovered in the process, h_r, often by burning byproducts and very often by burning extra coal in a boiler. It is safe to assume that all recovered heat can be used even though it is at a low temperature. If h_i' is the heating value of the coal fed to the plant (process plus boiler) and h_b' is the heating value of the byproducts leaving the plant (production minus utilization), the process train heat balance given by Eq. (4.17) can be extended to an overall plant balance

$$h_i' - h_p - h_b' = (h_{Ld} + h_{Lw} + h_{La}) + (h_f + h_s + h_m). \qquad (4.18)$$

The plant conversion efficiency defined by Eq. (4.16) is then written

$$\eta = \frac{h_p + h_b'}{h_i'}. \qquad (4.19)$$

All the energy supplied to drive the plant must ultimately be lost to the atmosphere. To investigate cooling water quantities it is necessary to find how much of the driving energy is lost (1) directly, (2) to condensers in acid gas removal regenerators, (3) to condensers on steam turbines, and (4) to interstage coolers on gas compressors. These categories encompass 85 percent or more of the losses but, as shown in the examples in the next chapters, an allowance should be made for unlisted additions. When fuel or coal is burnt, 10 to 15 percent of the heat will be lost directly to the atmosphere. The rest will go into steam or heating process streams. The heat in the steam used in acid gas removal regenerators is lost in the condenser, but an estimate must be made of how the energy in the rest of the heating system is lost. About 70 percent of the steam fed to turbines is lost in the turbine condensers. If the turbines are driving gas compressors, the other 30 percent is lost in the interstage coolers. If the turbines are driving electric generators or pumps for liquids or slurries, the other 30 percent is dissipated directly to the atmosphere by low temperature radiative and convective heat transfer from surfaces.

The above procedure has yielded an estimate of the various types of cooling loads in the plant. Judgement must now be used to decide how to divide these loads between wet and dry cooling. These decisions, and not the heat balance estimates, control the estimate of cooling water quantity. We have seen that process streams will be cooled to about 140°F by dry cooling and below that by wet cooling. In gas purification 90 percent or more of the total cooling load will be dry cooled, with the remaining 10 percent or less wet cooled. Where water is moderately priced the condensers on the industrial-type steam turbines will all be wet cooled. When water is expensive, however, (greater than about $0.80/10^3$ gal) or of limited availability, combined wet and dry cooling systems would be used. For such systems the average annual water consumption is between 10 and 25 percent of the all wet system. In general, so long as water is not very expensive (less than about $1.50/10^3$ gal) wet interstage cooling will be used on gas compressors. However, should water be very expensive or not available, a combined wet/dry system would be used to reduce the water consumption to about 10 percent of that for all wet cooling. Where wet cooling is used, about 1 lb of water will be evaporated for every 1,400 Btu dissipated. Makeup water can be assumed to be 10 percent more than the evaporated water, with the extra to compensate for blowdown.

Chapter 4. References

1. Leung, P. and Moore, R. E., "Water Consumption Determination for Steam Power Plant Cooling Tower: A Heat-and-Mass Balance Method," Paper No. 69-WA/PWR-3, ASME Winter Annual Meeting, Los Angeles, Nov. 1969.

2. Leung, P., "Evaporative and Dry-Type Cooling Towers and Their Application to Utility Systems," *Water Management by the Electric Power Industry* (E. F. Gloyna, H. H. Woodson, and H.R. Drew, eds.), pp. 106–116, Center for Research in Water Resources, the University of Texas at Austin, 1975.

3. Roffman, A. et. al., "The State of the Art of Saltwater Cooling Towers for Steam Electric Generating Plants," Report No. WASH 1244, Atomic Energy Comm., Washington, D.C., Feb. 1973.

4. Miliaras, E. S., *Power Plants with Air-Cooled Condensing Systems*. MIT Press, Cambridge, Mass., 1974.

5. Dickey, B. R., Grimmett, E. S., and Killian, D. C., "Waste Heat Disposal Using Fluid-Bed Wet-Dry Towers," *Water-1973, AIChE Symp. Series,* Vol. 70, No. 136, pp. 430–436, 1974.

6. Andeen, B. R., Glicksman, L. R., and Rohsenow, W. M., "Heat Rejection from Horizontal Tubes to Shallow Fluidized Beds," Report No. MIT-EL74-007, M.I.T. Energy Lab., M.I.T., Cambridge, Mass., 1974.

7. Barile, R. G., "Turbulent Bed Cooling Tower," Report No. EPA-660/2-75-027, Environmental Protection Agency, Corvallis, Oregon, 1975.

8. Robertson, M. W. and Glicksman, L. R., "Periodic Cooling Towers for Electric Power Plants," *Dry and Wet/Dry Cooling Towers for Power Plants* (R. L. Webb, ed.), ASME, New York, 1973.

9. Larinoff, M. W. and Forster, L. L., "Dry and Wet-Peaking Tower Cooling Systems for Power Plant Application," Paper No. 75-WA/PWR-2, ASME Winter Annual Meeting, Houston, Texas, Dec. 1975.

10. Malina, Jr., J. A. and Moseley, II, J. C., "Costs of Alternative Cooling Systems," *Water Management by the Electric Power Industry* (E. F. Gloyna, H. H. Woodson, and H. R. Drew, eds.), pp. 149–162, Center for Research and Water Resources, The University of Texas at Austin, 1975.

11. Hauser, L. G. and Oleson, K. A., "Comparison of Evaporative Losses in Various Condenser Cooling Water Systems," *Proc. American Power Conference,* Vol. 32, pp. 519–527, 1970.

12. Skamser, R., "Coal Gasification, Commercial Concepts, Gas Cost Guidelines," Report No. FE-1225-1, Energy Res. & Develop. Admin., Washington, D.C., Jan. 1976.

13. MacFarlane, J. A., Goodling, J. S., and Maples, G., "Rejection of Waste Heat from Power Plants Through Phased-Cooling," Paper No. 74-PWR-1, IEEE-ASME Joint Power Generation Conference, Miami, Fla., Sept. 1974.

14. Rohsenow, W. M. and Hartnett, J. P. (eds.), *Handbook of Heat Transfer.* McGraw-Hill, N.Y., 1973.

15. Goldstein, D. J. and Probstein, R. F., "Water Requirements for an Integrated SNG Plant and Mine Operation," *Symp. Proc.: Environmental Aspects of Fuel Conversion Technology II,* pp. 307–332, Report No. EPA-600/2-76-149, Environmental Protection Agency, Research Triangle Park, N.C., June 1976.

16. Dravo Corp., "Handbook of Gasifiers and Gas Treatment Systems," Report No. FE-1772-11, Energy Res. & Develop. Admin., Washington, D.C., Feb. 1976.

17. *Cooling Performance Curves.* Cooling Tower Institute, Houston, Texas, 1967.

18. *Kelly's Handbook of Crossflow Cooling Tower Performance.* Neil W. Kelly and Associates, Kansas City, Missouri, 1976.

19. Perry, R. H. and Chilton, C. H. (eds.), *Chemical Engineer's Handbook.* 5th Edition, McGraw-Hill, New York, 1973.

20. Gold, H., Goldstein, D. J., and Yung, D., "Effect of Water Treatment on the Comparative Costs of Evaporative and Dry Cooled Power Plants," Report No. COO-2580-1, Div. of Nuclear Res. & Application, Energy Res. & Develop. Admin., Washington, D.C., July 1976.

5
Gas Production

5.1 Gasification Technologies

In this chapter we assess the quantity and quality of water fed to and evolved from various processes to convert coal to gas, as well as the quantity of water evaporated for cooling. Cooling water is usually the largest use. However, knowledge of the input and output streams, termed "process requirements," is important in the design of a water treatment plant necessary to any estimate of water needs, because the quality of process water is at both extremes. Most of the water is fed to the process as steam and so must be of the highest purity. Most of the water evolved from the majority of processes is an extremely dirty condensate which has contacted coal, tar, and gas. To evaluate these water streams, the general route of gas production and the details of the process technology or a description of the process type must be known.

A simplified block diagram of the important sections and water streams of a coal gasification plant for pipeline gas production is shown in Fig. 5-1. Not all the sections are needed for every process. Only one of the three sections marked oxygen plant, hydrogen plant, and indirect heat is used in any single process, and in some special molten media gasifiers none may be required.

To produce a thousand pounds of methane from coal with a carbon-to-hydrogen weight ratio of 15, the minimum theoretical amount of water is 216 gallons (Table 2.1). The gas industry uses 379.7 standard cubic feet of gas per pound mole of gas or 23.73 scf per pound of methane. It follows that a methane output of 250×10^6 scf/day, appropriate to a standard size plant, contains hydrogen equivalent to 1,580 gal/min of water. In fact, more water than is needed for the chemical reaction is usually added to the gasifiers to control the temperature and moderate the reaction.

The surplus water comes out of the process in several places. Some water is condensed from the raw gas leaving the gasifiers. This water is called process condensate, and it is the surplus water actually removed from the process. In some processes the raw gas is scrubbed with water. This cools the gas and transfers tar, dust, and soluble organic materials from the gas to the water. Cooling also causes condensation of water vapor that left the gasifier. The scrubbing water is circulated, and only the surplus water from condensation is removed from the process.

Although many coals contain a high percentage of moisture, this water is not usually available to enter into chemical reaction. Often the coal must be dried before it is fed to the gasifer. Even if coal is not predried it is usually fed near the top of the reactor, where the water vaporizes and leaves before it has time to

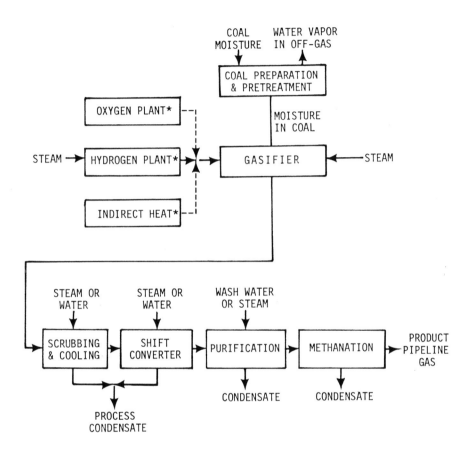

Figure 5-1
Water streams in a plant to produce pipeline gas from coal.

react. A wet coal will usually increase the amount of process condensate and not
reduce the steam requirement to the gasifier.

Sometimes additional cooling causes water to condense in the gas purification
liquid. Such water must then be boiled out so that the gas purification liquid can
be recycled. This water is shown in Fig. 5-1 as leaving the purification stage.
Water is formed during the methanation stage and is condensed and knocked out
of the product gas. This water is shown in Fig. 5-1 as leaving the methanator.

The stages in a plant to produce power gas (low-Btu gas) from coal are shown

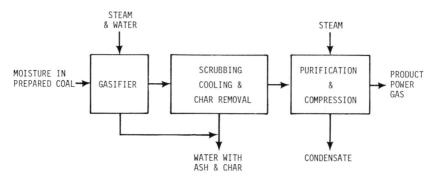

Figure 5-2
Water streams in a plant to produce power gas from coal.

in Fig. 5-2. Most of the comments given above for the production of pipeline gas apply to the production of power gas.

A large number of gasifier systems are commercially available or have the potential to become so. Twenty-two such gasifiers are discussed in a handbook of gasifier systems compiled by the Dravo Corporation.[1] A list and description of sixty-five gasification processes investigated prior to 1965 and processes currently under investigation may also be found in Ref. 2.

In the Dravo handbook the gasifiers are classified into four categories by the manner in which the coal or char is handled in the gasifier vessel. The categories of gasifiers are entrained flow, fixed bed, fluidized bed, and molten media. *Entrained flow gasifiers* all use pulverized coal (and char) pulverized typically to a size that allows 70 to 90 percent of the feed to pass through a 200 mesh screen (.003 inch opening). Most often steam with oxygen or air is used to entrain the pulverized coal, which is injected through nozzles into the gasifier burner. Or, hot product gas may be used to entrain the coal and at the same time gasify it. In *fixed bed gasifiers* coal nuggets, of between $\frac{1}{4}$ and $1\frac{1}{2}$ inches, a size range depending on the process, are fed into the top of the gasifier. The burning coal bed in the vessel sits on a grate or series of grates, which are rotated or agitated so the ash can fall to the bottom for removal. *Fluidized bed gasifiers* are fed with pulverized or crushed coal, depending on the process. The finest pulverized coal has 60 to 70 percent passing through a 200 mesh screen. The coarsest crushed coal is about $\frac{1}{4}$ inch. In fluidization, gas lifts the coal, thereby improving the mixing characteristics and bringing a more rapid approach to equilibrium. The fluidization in the vessel or stages of the vessel may take place where devolatilization, combustion, or gasification occurs. The lifting fluid may be a feed or product gas. In *molten media gasifiers,* coal is dispersed into a molten carrier, which generally takes part in the gasification reaction and acts as a heat source. The coal size depends on how it is fed into the molten media. A range of sizes less than ¼ inch will generally be acceptable, although if the coal is to be blown into the melt, then it must be pulverized as in any entrained flow system.

With some exceptions, the reaction temperature broadly defines the characteristics of the product gas from the gasifier. This temperature varies from about 1500° to 3500°F, and each type of gasifier covers a specific range. The exception to this is the molten media gasifier whose temperature and, to some extent, characteristics of operation are determined by the melt employed, for example, slag or a salt. For this reason, the list of general gasifier characteristics in Table 5.1 does not include molten media gasifiers. Most of their characteristics are the same as those for the other gasifiers in the same temperature range.

The most important feature in Table 5.1 is that entrained flow gasifiers operate at the highest temperature under conditions where the coal slags, while fluidized bed gasifiers operate at the lowest temperature. Low temperature operation generally favors methane formation. At high temperatures the CO-to-CO_2 ratio is high and the conversion of the carbon in the coal is high. This may be seen in the fact that the H_2-to-CO ratio is lowest in entrained flow gasifiers, an unfavorable condition for pipeline gas production, since more shift conversion is required. On the other hand, the cleanliness of the gas reflects the extent of the high coal conversion. The high reaction temperature ensures that no tars or heavy hydrocarbons are in the product gas, nor carbon in the slag. This high utilization of the carbon in the coal should not be confused with the overall conversion efficiency, which is lowest in the entrained flow gasifier when used for pipeline gas production.

The lower overall conversion efficiency in entrained flow gasifiers is a consequence of the fact that the least methane is formed in this type of gasifier, and thus less of the heat of methanation can be utilized. The exclusion of hydrogen-blown gasifier characteristics from Table 5.1 should be noted. This method of gasification is more efficient than oxygen-blown processes precisely because more of the methanation heat is utilized. However, the methanation temperature usually must be maintained below 900°F to protect the catalyst, and this low temperature energy cannot be used in the gasifier. The heat generated during hydrogasification can be used in the endothermic gasification reaction (Eq. 3.6) so that the oxygen requirement is also reduced. All reactor schemes therefore try to maximize the ratio of methane directly produced to the total methane. Of course, the higher the conversion efficiency the lower the cooling water require-

Table 5.1
Gasifier characteristics.

Type	Temperature (1,500°−3,500°F)	CH$_4$ (0–15 mol%)	H$_2$/CO† (0.4–2.5)	Conversion Eff. (65–80%)	Product Gas Cleanliness
Entrained Flow	High	Generally Little	Low–Medium	Lower	Clean
Fixed Bed	Intermediate	Medium	Medium–High	Upper	Dirty
Fluidized Bed*	Low	Can Be High	Medium–High	Upper	Intermediate

*Excluding hydrogen-blown gasifiers.
†Oxygen-blown and excluding CO$_2$ Acceptor.

ments will be. The difficulties in hydrogasification lie in the agglomerating properties of most American coals in a high temperature, high pressure hydrogen atmosphere.

Judging from the characteristics in Table 5.1, entrained flow gasifiers seem to be the more likely candidates for the production of power gas and industrial gas, and fixed bed and fluidized bed systems the more appropriate choice for pipeline gas production. From the point of view of water cleanup, the dirtiest process condensate is usually produced in fixed bed systems, where relatively large quantities of tar, oil, phenols, and other organics are condensed out.

In Tables 5.2 and 5.3 more details are given on the operating characteristics and gas products of 19 commercially available gasifiers or gasifiers with commercial potential.[1] As a rule, in any given category, the lower the reaction temperature is, the higher the operating pressure. Although in a medium-Btu product, a high H$_2$-to-CO ratio is desirable to minimize the shift conversion losses in pipeline gas production, a gas at atmospheric pressure is generally not desirable because compression energy is then required for methanation and for delivery of the product.

Another gasification technology receiving some attention is underground or *in situ* gasification.[3,4] In this method bore holes are drilled to the coal seam, and air or oxygen, with or without steam, is injected into the seam through the wells. The coal is ignited and combustion proceeds between the wells through permeable pathways in the coal seam. The gases that are formed are brought to the surface through drill holes. The processing of the gas recovered at atmospheric pressure is essentially the same as that from a corresponding reactor vessel, except that dust must be removed. However, the technology is still relatively

Table 5.2
Entrained flow and fixed bed gasifiers which may be oxygen- or air-blown.

Type and Name	Comb. Zone Temp. (°F)	Pressure (psig)	Typical Mole Ratio H_2/CO	CO/CO_2
Entrained Flow				
Koppers-Totzek	3,500	~atm	0.69	5.3
Babcock & Wilcox	3,400	atm–300	0.43	13.0
Combustion Engineering	3,000	atm	0.77[a]	3.2[a]
Bi-Gas	2,700–3,000*	500–1,500	1.1	1.4
Foster Wheeler	2,500–2,800	350	0.49[a]	8.9[a]
Texaco	> ash fusion	300–1,200	1.0	1.8
Fixed Bed				
ERDA Stirred Bed	2,400–2,500	atm–285	0.76[a]	2.3[a]
Wellman-Galusha	2,400	~atm	0.52[a]	8.4[a]
Woodhall Duckham/ Gas Integrale	2,200**	atm	1.0	2.1
Lurgi	1,800–2,500†	350–450	2.3	0.54

*Second stage ~ 2,200°F.
**Dependent on coal type.
†Gasification zone, 1,200°–1,500°F. Also operable slagging.[5]
[a]Air-blown, otherwise oxygen-blown.

undeveloped, difficulties still exist in quality and quantity control, and some environmental problems are anticipated. This method will therefore not be considered further other than to note that the criteria for processing applied to other gasifiers are appropriate here.

5.2 Hydrogen Balance and the Synthane Process

To define the various process water streams in any gasification process, an integrated design for the complete plant as conceived for a commercial operation must be given. This design is distinguished from pilot plant or demonstration plant configurations which generally do not incorporate all of the features or operations included in a full-scale system. At the heart of the design and the most unique part of the plant is the gasifier itself. The output of the gasifier as a function of the operating and feed conditions must be specified on the basis of

Table 5.3
Fluidized bed and molten media gasifiers.

Type and Name	Maximum Temp. (°F)	Pressure (psig)	Typical Mole Ratio	
			H_2/CO	CO/CO_2
Direct Heat (O_2, Air)				
Westinghouse	2,100	130–200	0.74[a]	2.1[a]
Synthane	1,800	1,000	2.5	0.37
Winkler	1,500–1,800*	atm	0.73	1.4
Hydrogasifiers (O_2)				
Hygas**	1,850	1,160–1,175	1.3	0.97
Hydrane	1,800	1,000	3.6	3.5
Indirect Heat (Air)				
Battelle/Carbide Agglomerating Ash†	2,000–2,100	100	Wide Range	
CO_2 Acceptor††	1,850	150	3.8	1.7
Molten Media (O_2, Air)				
Otto-Rummel	2,700–3,100	atm–360	0.57	3.8
AI Molten Salt	1,800	atm–280	0.44[a]	8.5[a]

*Range for low to high rank coals.
**Hydrogasifier stage II 1,700°–1,800°F, stage I 1,200°–1,300°F.
†Gasifier, 1,600°–1,800°F.
††Gasifier, 1,500°F.
[a]Air-blown.

prior knowledge or experiment. It cannot be specified on purely theoretical grounds, because the gasifier is not in chemical equilibrium and the extent of the disequilibrium can normally only be determined empirically.

In this section a detailed procedure is given for performing a material balance for a specific plant design. The material balance is then presented as a total hydrogen balance expressed in weights of water equivalent. This defines the quantities of process water streams, both vapor and liquid. In principle, any material balance should close, although it may not close in practice, depending on minor details of the process design. To illustrate the details of the procedure, the Synthane process for the production of pipeline gas was chosen.

The Synthane process is oriented to the production of high-Btu pipeline gas and is being developed by the Pittsburgh Energy Research Center of the U.S. Department of Energy. The gasifier shown in Fig. 5-3 is a vertical, high pressure

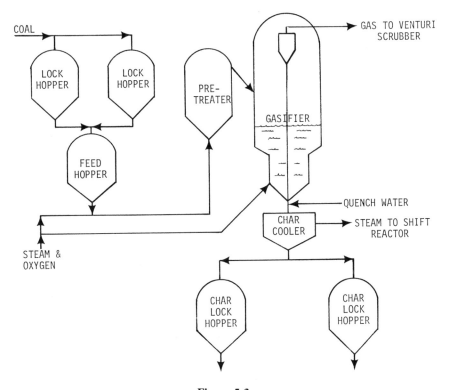

Figure 5-3
Flow diagram for Synthane gasifier.

fluidized bed reactor. A pilot plant capable of handling 72 tons/day of coal or lignite at 1000 psig is located at Bruceton, Pennsylvania. Shift conversion and methanation are part of the plant's operation. The gasifier is 101 ft high and has an inside diameter of 5 ft.[1] Since the throughput goes as the square of the diameter, a commercial size unit to handle 1150 tons/day of coal would be about 20 ft in diameter. About 20 such units would be required for a plant designed to produce 250×10^6 scf/day of pipeline gas.

Pulverized and dried coal that can pass through 20 mesh screens (.033 inch opening) is fed to a high pressure pretreater through lock hoppers. The purpose of the pretreater is to carry out a mild oxidation of the coal to eliminate the tendency in certain coals to cake when heated in a hydrogen atmosphere. Although mild oxidation with air or oxygen can prevent coal from agglomerating, it also adversely reduces the coal's reactivity for methane formation and consequently the process conversion efficiency. The pretreater here is a fluid bed reactor in which about 12.5 percent of the total steam and oxygen fed to the gasifier reacts with the coal at about 1000 psig and 750° to 800°F. If the coal is non-caking, it is simply passed through the pretreater. The coal overflows from the pretreater into the gasifier, falls freely some 20 feet, and enters the fluidized bed maintained in the lower section. Steam and oxygen enter through the bottom. A temperature gradient of 1400° to 1800°F is maintained from the top to the bottom of the fluidized bed. The gas produced flows up the gasifier, devolatilizing the coal which is falling countercurrently. It then passes through an internal cyclone that removes particles larger than 50 microns and returns them to the fluidized bed.

About one third of the carbon fed to the gasifier is not gasified but instead is released as char and tar. (The char is removed from the bottom.) In the conceptual design presented below, part of the char is burnt to generate steam, to provide power for the plant. The remainder of the char is presumed to be sold.

Figure 5-4 is a simplified flow diagram of a plant designed to produce 250×10^6 scf/day of pipeline gas by the Synthane process. The product gas has a heating value of 940 Btu/scf (92.3 percent methane, 1.8 percent hydrogen). The material flows into and out of the gasifier have been taken from Ref. 6. All other streams have been calculated. The coal chosen for the example is a Wyoming subbituminous coal. Its characteristics as received from the mine and as fed to the gasifier after drying are shown in Table 5.4. The feed rate of 19,260 tons/day of dried coal to the gasifier corresponds to an as-received rate of 23,016 tons/day.

The simplest method by which to illustrate the material balance calculations is to follow the flow diagram of Fig. 5-4 and discuss each stream in turn. Streams ① to ③ are the input quantities. The specified composition of the output from the gasifier, represented by stream ④, and the calculated compositions for the other gas streams leaving the major sections of the plant are given in Table 5.5. The water streams in the plant are given in Table 5.6.

Stream ⑤ is the foul water that comes out from the scrubbing of the gas leaving the gasifier. It is the difference between the amount of water vapor leaving in the gasifier stream and the amount of water vapor saturating the gas at 267°F. Stream ⑥ is the steam generated from the foul water used to cool down the char from 1700° to 800°F. This steam is used in the shift reaction. Stream ⑦ is the additional steam produced from boiler feed water needed for the shift reaction. The quantity is determined by the amount required for the shift conversion minus the quantity in stream ⑥.

In the shift converter, CO is reacted over a catalyst with steam to produce H_2 and CO_2 (Eq. 3.8). Around the shift converter, the C, H_2, and O going in and out are balanced. Assuming that the CH_4 and C_2H_6 in the incoming gas are unaf-

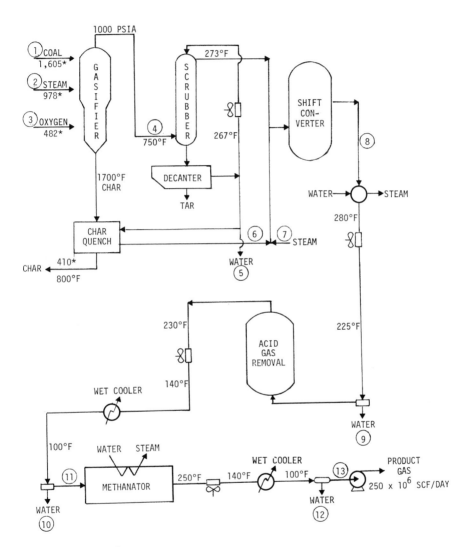

*FLOW RATES IN 10^3 LB/HR

Figure 5-4
Flow diagram for Synthane process for the production of 250×10^6 scf/day of
pipeline gas.

Table 5.4
Coal used in the Synthane example.
Wyoming subbituminous, Wyodak seam.

	Weight Percent	
	As-Received	Fed to Gasifier
C	54.0	64.5
H	3.5	4.1
N	0.8	1.0
S	0.6	0.8
O	14.1	16.8
Ash	7.1	8.5
Moisture	19.9	4.3
Total	100	100
HHV (Btu/lb)	8,905	10,640

Table 5.5
Process stream compositions from major sections of Synthane plant in 10^3 moles/hr.

	Gasification ④	Shift ⑧	Purification ⑪	Methanation ⑬
CO	16.70	7.82	7.82	0.03
CO_2	25.89	34.77	1.02	1.04
H_2	16.03	24.91	24.91	0.53
CH_4	15.24	15.24	15.22	25.22
C_2H_6	1.12	1.12	1.12	—
H_2O	36.43	9.38	0.05	0.03

Table 5.6
Water streams in Synthane plant.

Stream	H_2O (10^3 moles/hr)	Stream	H_2O (10^3 moles/hr)
⑤	28.68	⑨	7.58
⑥	4.27	⑩	1.09
⑦	10.51	⑫	7.76

fected by the reaction, the concentrations of CO, CO_2, H_2, and H_2O per unit time leaving the gasifier must be determined. The total steam input rate comprising streams ⑥ and ⑦ is also unknown. Three of the five equations needed to determine the five unknowns are given by the material balance relations for C, H_2, and O. The fourth, the law of mass action defining the equilibrium concentrations for the reversible shift reaction Eq. (3.8), may be written

$$\frac{[CO_2] \, [H_2]}{[CO] \, [H_2O]} = K, \tag{5.1}$$

where K is the equilibrium constant and the brackets denote molar concentrations. Although the mean temperature of the gas leaving the shift converter is 700°F, an equilibrium constant of 1.8, corresponding to 750°F,[7] is applied as an approximate way of accounting for the fact that equilibrium is not quite attained and the conversion thus not quite complete. The last relation is provided by the condition that the molar ratio of H_2 to CO should be at least 3-to-1, as the subsequent methanation step requires. In the present design the value for this ratio was taken to be 3.18. The calculated compositions for the stream ⑧ leaving the shift converter and the amount of boiler steam required in stream ⑦ are given in Tables 5.5 and 5.6, respectively.

Following the shift conversion, the gas stream is cooled down to 225°F and the difference in water vapor between that leaving the shift converter and the amount saturating the gas at 225°F is condensed out and removed as stream ⑨. The gas then enters the gas purification stage. In the Benfield hot potassium carbonate system, 98 percent of the CO_2 is assumed to be removed. The gas is then cooled down to 100°F and the difference in water vapor between that leaving the absorber and the amount saturating the gas at 100°F is condensed out as stream ⑩. The resulting gas stream ⑪ leaving the char tower is the purified gas whose composition is given in Table 5.5.

The last major section of the plant is the methanator in which the chemistry follows the reaction of Eq. (3.9). It is assumed that all of the CO is removed and the methanation reaction is carried to completion. The surplus water of reaction generated in the methanator is a very clean water, and the excess over saturation, which is condensed out at 100°F, is stream ⑫.

On the basis of the information given above, the hydrogen fed to the plant in the coal, in the coal moisture, and in the steam to the gasifier and shift converter was balanced against the hydrogen leaving in the various streams. These results, expressed in weights of water equivalent, are given in Table 5.7. The balance in this case is quite close, in part a result of adjustments in the design itself. Using

Table 5.7
Water equivalent hydrogen balance for Synthane plant producing 250×10^6 scf/day
of 940 Btu/scf pipeline gas from 23,016 tons/day of as-received
Wyoming subbituminous coal.

	10^3 lb/hr	gal/10^6 Btu*
In		
Moisture in coal fed to gasifier	69	0.85
Water equiv. of H_2 in coal	604	7.41
Steam to gasifier and shift converter	1,167	14.31
Total	1,840	22.57
Out		
Foul condensate from scrubbing	516	6.33
Condensate from shift conversion	138	1.69
Condensate from acid gas removal	20	0.25
Clean methanation condensate	140	1.72
Water equiv. of H_2 in byproducts and lost gas	87	1.07
Water equiv. of H_2 in product gas	920	11.28
Total	1,821	22.34

*In the product gas.

the results of Table 5.7, the quantities of actual water streams flowing in and out
of the process train in Fig. 5-1 can now be identified.

5.3 Hydrogen Balances for Other Processes

In this section material balances are given for a number of other gasification
processes embodying the types classified in Section 5.1, to provide a representative picture of the process water streams in coal gasification. Table 5.8 lists the
gasifiers and products to be considered.

Table 5.8
Representative gasification processes.

Type	Name	Product
Fluid bed, direct heat	Synthane	Pipeline gas
Fluid bed, hydrogasifier	Hygas	Pipeline gas
Fluid bed, indirect heat	CO_2 Acceptor	Pipeline gas
Fixed bed	Lurgi	Pipeline and power gas
Entrained flow	Koppers-Totzek	Power gas

Hygas

The Hygas process, the heart of which is a fluid bed hydrogasifier, has been under development by the Institute of Gas Technology since 1945. The process has advanced to the large pilot plant stage. The facility, located in Chicago, is capable of handling 80 tons/day of coal at about 1200 psig. Design and construction of a demonstration plant under the auspices of the U.S. Department of Energy is planned. In the integrated Hygas process, the hydrogen-rich stream is produced by the gasification of residual char from two stages of hydrogasification. There are three processes for char gasification, but only the one using steam and oxygen will be described here. The pilot plant gasifier is a reaction vessel 132 ft high with an inside diameter of about 5.5 ft and five internally connected gas-solids contacting stages. A sketch of the reactor, its stages and operation are shown in Fig. 5-5.

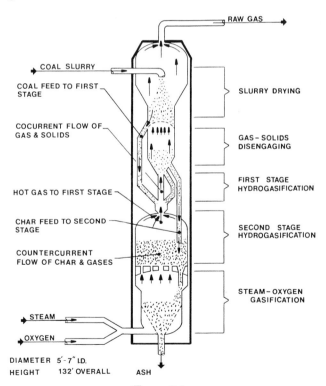

Figure 5-5
Hygas pilot plant hydrogasification reactor.

The coal is crushed to a feed having a maximum size of 10 mesh (.079 inch opening) a minimum of fines of less than 100 mesh (.006 inch opening), and is dried to a 2 percent moisture content. If the coal is an agglomerating type, it may need pretreatment, which involves air oxidation in a separate vessel at 750° to 800°F and atmospheric pressure. The coal is then slurried to a 50 percent by weight solids concentration, with recycled light oil produced as a byproduct downstream in the process. The slurry is pumped to the gasifier operating pres-

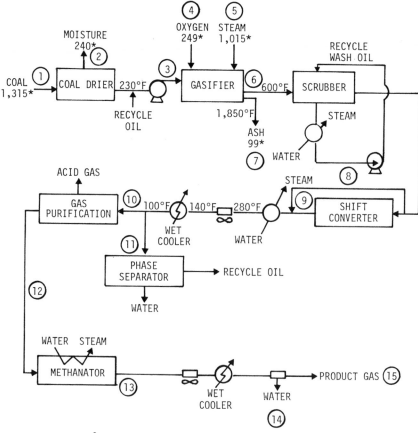

*FLOW RATES IN 10^3 LB/HR

Figure 5-6
Flow diagram for Hygas process for the production of 250×10^6 scf/day of pipeline gas.

sure of around 1200 psig and if necessary heated in an external heater to 200°F. It is then injected into the fluidized top section as a spray. Sensible heat in the gaseous products efficiently vaporizes the light oil and leaves dry coal for feeding to the second section.

In the dilute-phase second section (the first stage of hydrogasification), hot gases at about 1700° to 1800°F, rising from the second stage of hydrogasification, react with the coal in concurrent flow. In this stage about 20 percent of the coal is converted to methane at about 1200° to 1300°F. In the second stage of hydrogasification, hydrogen reacts exothermically with the char to produce methane, and steam reacts endothermically with char to produce CO and more hydrogen. About 25 percent more of the coal is converted in this stage. The hot char descends to the final dense-phase fluidized bed where the hydrogen-rich gas is produced in the presence of steam and oxygen.

Figure 5-6 shows a simplified flow diagram of a plant designed to produce 250×10^6 scf/day of pipeline gas by the Hygas process. The product gas has a heating value of 969 Btu/scf (95.4 percent methane, 1.7 percent hydrogen). The material flows into and out of the gasifier were supplied by the Institute of Gas Technology; all other streams were calculated. The coal in this example is a Wyoming subbituminous; its composition and that of the residual ash are shown in Table 5.9. The feed rate of as-received coal to the gasifier is 15,774 tons/day. An additional 2,550 tons/day is fed to the boilers to generate steam. The compo-

Table 5.9
Coal and ash in the Hygas example. Wyoming subbituminous, Wyodak seam.

	Weight Percent		
	As-Received	Fed to Gasifier	Ash Residue
C	54.2	66.31	17.80
H	4.0	4.93	0.19
N	0.8	0.94	2.01
S	0.6	0.74	0.38
O	14.5	17.74	—
Ash	6.0	7.34	79.62
Moisture	19.9	2.00	—
Total	100	100	100
HHV(Btu/lb)	9,264	11,334	2,722

Table 5.10
Gasifier and product stream compositions in Hygas plant.

	Slurry Oil ③ (wt.%)	Gasifier Off-Gas ⑥ (10^3 moles/hr)	Product Gas ⑮ (10^3 moles/hr)
CO	—	20.55	0.03
CO_2	—	19.32	0.69
H_2	—	25.21	0.48
CH_4	—	13.77	26.18
C_2H_6	—	1.04	—
C_6H_6	10	1.53	—
C_7H_8	85	10.70	—
H_2O	—	25.81	0.03
Other	5*	0.76	0.04

*C_8, C_9, C_{10} aromatics.

sitions of the slurry oil, gasifier off-gas, and product gas are given in Table 5.10.

The raw off-gas from the gasifier contains materials other than fuel gas species. For example, the oil made in this type of gasifier is expected to be approximately 85 percent toluene and 15 percent benzene with a small quantity of phenol. It is significantly lighter than the oil produced in most other gasification systems. The oil made about equals the oil lost in purification or left in the product gas. To cool the gasifier product gas and remove the particulate matter, the gas is quenched with the oil to about 400°F.

A portion of the gas undergoes shift reaction at an equilibrium temperature of 750°F to adjust the H_2-to-CO ratio for the downstream methanation reactor. The shifted gas is cooled to 100°F to ensure condensation of the oil. The water condensed at this point is used to remove the soluble species from the gas. A physical solvent-based system, such as the Selexol process,[1] removes acid gas and provides a treated gas of sufficient purity so that only a nominal sulfur guard is required prior to methanation. The gas purification process is assumed to reduce CO_2 to 1 percent and completely absorb all other acid gases. Based on the recommendations of the Institute of Gas Technology, a loss of 0.5 percent of H_2 and CO, a 1 percent loss of CH_4 and a 25 percent loss of C_2H_6 are assumed to occur in gas purification. The resulting hydrogen balance defining the quantities of the water streams flowing into and out of the process train is given in Table 5.11.

Table 5.11
Water equivalent hydrogen balance for Hygas plant producing 250×10^6 scf/day of 969 Btu/scf pipeline gas from 18,324 tons/day of as-received Wyoming subbituminous coal.

	10^3 lb/hr	gal/10^6 Btu*
In		
Moisture in coal fed to gasifier	21	0.25
Water equiv. of H_2 in coal	477	5.67
Steam to gasifier	1,015	12.07
Total	1,513	17.99
Out		
Condensate in waste heat recovery phase separator	294	3.50
Water equiv. of H_2 from acid gas removal	60	0.71
Clean methanation condensate	201	2.39
Water equiv. of H_2 in product gas	951	11.31
Total	1,506	17.91

*In the product gas.

CO_2 Acceptor

The CO_2 Acceptor fluidized bed process to convert lignite or subbituminous coal to pipeline gas is being developed by Conoco Coal Development Co.[1,8] The mineral dolomite ($MgO \cdot CaCO_3$) plays a central role in the process. When heated (calcined) the dolomite releases CO_2. The hot calcined dolomite ($MgO \cdot CaO$) is added to the gasifier where it reacts with CO_2 in the gas formed from the coal. The calcined dolomite is a sensible heat carrier and an "acceptor" material, hence the name CO_2 Acceptor process. In addition to the sensible heat, an exothermal heat release also accompanies the acceptor reaction

$$MgO \cdot CaO + CO_2 \rightarrow MgO \cdot CaCO_3. \qquad (5.2)$$

The heat release and the indirect sensible heat provide the process heat for the gasification reaction. A 40 ton/day pilot plant is located in Rapid City, South Dakota.

Raw lignite crushed to a feed having a maximum size of 8 mesh (.094 inch opening) and a minimum of fines less than 100 mesh (.006 inch opening) is lifted with hot flue gas to a preheater. The lignite, originally about 35 percent by weight moisture, enters the preheater with only about 5 percent. The preheater operates at atmospheric pressure and 400° to 500°F. Pyrolysis of the coal does not occur. The residence time in the pilot plant preheater is 16 hours.

The preheated lignite is fed to the bottom of the char phase of the fluid bed gasifier via lock hoppers, as shown in Fig. 5-7. Steam introduced into the gasifier reacts with the lignite. The gasifier and regenerator operate at 150 psig. The hot acceptor, from the regenerator, showers through the lignite bed, absorbing CO_2, driving the shift reaction (Eq. 3.8) to completion, and enriching the gas leaving the gasifier with hydrogen. The absorption of CO_2 also supplies heat to maintain the gasifier in the range of 1480° to 1520°F.

Acceptor material accumulates at the bottom of the gasifier from which it is eventually allowed to flow out. Char, representing about 33 percent of the carbon in the feed coal, is allowed to flow from the top of the gasifier. Separate exit points for acceptor and char allow a concentration ratio of char to acceptor in the

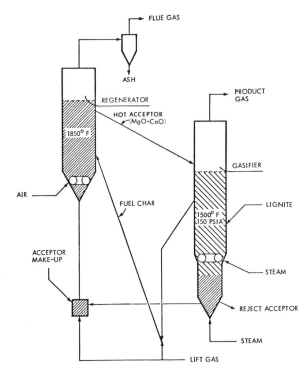

Figure 5-7
Flow diagram for CO_2 Acceptor gasifier.

gasifier different than the ratio of the circulation rates. In the pilot plant char is withdrawn at 600 lb/hr and acceptor is withdrawn at 1100 lb/hr, but the ratio of char to acceptor in the gasifier is much higher than 6-to-11. The char residence time is greater than the residence time of acceptor in the gasifier and sufficiently long that phenols, tars, and oils are not formed.

In the pilot plant, acceptor is lifted to the regenerator by the flue gas. Fresh acceptor is added and used acceptor is withdrawn at a rate required to maintain the activity of the acceptor. In the pilot plant, char is lifted to the regenerator with nitrogen. In the regenerator, air is added to burn the char and raise the temperature to between 1840° and 1860°F. This reverses the acceptor reaction (Eq. 5.2) and drives off CO_2. Flue gas leaves the regenerator with entrained ash, which is separated out in cyclones and released via lock hoppers. The ash contains less than 10 percent carbon. About 10 to 20 percent of the sulfur in the feed coal leaves the gasifier as H_2S, and the rest of the sulfur is converted to CaS. Because the H_2-to-CO ratio exceeds 3-to-1, a shift reactor is not required.

In Table 5.12 a water equivalent hydrogen balance is presented, scaled to 250×10^6 scf/day of pipeline gas from a conceptual design for 260×10^6 scf/day.[8] The coal is a North Dakota lignite with a 34 percent moisture content and a

Table 5.12
Water equivalent hydrogen balance for CO_2 Acceptor plant producing 250×10^6 scf/day of 953 Btu/scf pipeline gas from 26,983 tons/day of as-received North Dakota lignite.*

	10^3 lb/hr	gal/10^6 Btu†
In		
Water equiv. of H_2 in preheated lignite	462	5.6
Steam to gasifier	1,035	12.5
Total	1,497	18.1
Out		
Water vapor in flue gas from regenerator	23	0.3
Condensate from cooling and spraying	155	1.9
Condensate from purification	115	1.4
Clean methanation condensate	256	3.1
Water equiv. of H_2 from purification (H_2S)	1	—
Water equiv. of H_2 in product gas	918	11.1
Total	1,468	17.8

*33.7 wt. % moisture, moisture-free heating value 11,120 Btu/lb.
†In the product gas.

heating value when dried of 11,120 Btu/lb. The feed rate of the as-received coal is 26,983 tons/day.

Lurgi

The Lurgi process is a commercially proven and available high pressure gasification process for manufacturing fuel gas. At least four full-sized commercial plants have been designed in the United States to utilize the Lurgi process to produce pipeline gas.

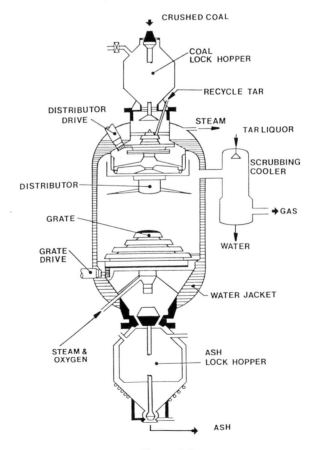

Figure 5-8
Lurgi pressure gasifier.

The Lurgi gasifier shown in Fig. 5-8 consists of several vertically stacked chambers. From top to bottom they are: the coal bunker, coal lock chamber, water-jacketed gas producer chamber, ash lock chamber, and ash quench chamber. Coal crushed and screened to $\frac{1}{8}$ in. $\times$ $1\frac{1}{2}$ in. flows through the lock chamber to the pressurized reactor maintained at 350 to 450 psig. The gasifier cannot accept fines; these must be disposed of or sold. Steam and oxygen are introduced through a revolving grate at the bottom of the reactor.

The moving bed of coal, which is the volume between the inlet and outlet grates, has several distinct zones. Devolatilization, including the vaporization of coal moisture, occurs at the top, and gasification begins lower down where the temperature reaches 1150° to 1400°F. The minimum residence time of coal at the desired temperature level of 1400° to 1600°F is about one hour. The bottom of the coal bed is the combustion zone in which about 14 percent of the coal fed to the gasifier is burned with oxygen to supply heat for the endothermic gasification reaction (Eq. 3.6). Solid ash is removed through the ash lock chamber at the bottom.

The crude gas leaves the gasifier at temperatures between 700° and 1100°F,

Table 5.13
Water equivalent hydrogen balance for Lurgi plant producing 250×10^6 scf/day of 954 Btu/scf pipeline gas from 20,192 tons/day of as-received Navajo subbituminous coal.*

	10^3 lb/hr	gal/10^6 Btu†
In		
Moisture in as-received coal	273	3.3
Water equiv. of H_2 in coal	545	6.6
Steam to gasifier	1,548	18.7
Boiler feed water to purification	87	1.1
Total	2,453	29.7
Out		
Dirty condensate from scrubber, shift reactor and purification	1,084	13.1
Clean methanation condensate	274	3.3
Water equiv. of H_2 in naptha and byproducts	206	2.5
Water equiv. of H_2 in product gas	943	11.5
Total	2,507	30.4

*16.3 wt. % moisture, as-received heating value 8,664 Btu/lb.
†In the product gas.

depending upon the type of coal. The gas contains carbonization products such as tar, oil, naptha, phenols, etc., and traces of coal and ash dust. Quenching with circulating water moves the contaminants from the gas into the water. A portion of the cooled crude gas goes to a shift conversion reactor where CO and H_2O are catalytically converted to H_2 and CO_2 so that the H_2-to-CO ratio of the mixed gas is adjusted to greater than 3, to be suitable for the methanation step. After shift conversion the acid gases are usually removed by the Rectisol process,[1] although other acid gas removal processes are feasible. Finally, the purified gas passes through the methanation stage, consisting of fixed bed reactors with pelleted, reduced nickel-type catalysts.

In Table 5.13 the hydrogen balance scaled to 250×10^6 scf/day is presented for the proposed El Paso gasification plant to be located in the "Four Corners" area of New Mexico.[9-11] The coal is a Navajo subbituminous coal with a 16 percent moisture content, about a 25 percent ash content, and a heating value of 8,664 Btu/lb. The feed rate of as-received coal is 20,192 tons/day. The most important point to be observed from the balance is that undried coal is fed directly to the gasifier; however, the moisture in the coal does not enter into the reaction but is evolved as a dirty condensate that must be cleaned.

Lurgi (Power Gas)

The Lurgi gasifier can also be air blown to yield a power gas. Here, brief mention will be given of recent developments by the British Gas Corporation and Lurgi in running the gasifier in a slagging mode for the production of a medium-Btu power gas of around 370 Btu/scf.[5]

The Lurgi gasifier for pipeline gas production uses about 7 moles steam/mole oxygen with Navajo coal; 45 to 50 percent of the steam decomposes. An air-blown Lurgi gasifier for utility fuel gas production with the same coal might typically use 3.5 to 4 moles steam/mole oxygen in air, and in these conditions 45 percent of the steam decomposes. Nitrogen in the air moderates the gasifier temperature as surplus steam does in an oxygen-blown gasifier. In the slagging mode, the oxygen-to-carbon ratio is slightly increased, but the ratio of steam to oxygen is reduced to about 1.5 and all of the steam decomposes. Only the coal moisture appears in the gas. This greatly reduces the steam requirement and hence the size of the water treatment plant. Also, when the gas is cooled to remove tars and oils, very little water condenses out. In the plants discussed in this chapter, condensing water is more than half of the cooling load. With little water to condense, irreversible loss on cooling is small so that slagging gasifiers have a higher cold gas efficiency than the usual Lurgi gasifier. The experimental slagging gasifier has been run at more than four times the throughput of the normal gasifier.[5] There is no doubt that the slagging gasifier will prove most useful. Under auspices of the U.S. Department of Energy, the Conoco Coal Development Co. has been contracted to build a demonstration plant to produce 58×10^6 scf/day of pipeline gas using a slagging Lurgi gasifier.

Koppers-Totzek

The Koppers-Totzek gasifier is a commercially established high temperature entrained flow gasifier capable of partially oxidizing a wide variety of feed stocks.[1] The gasifier operates slightly above atmospheric pressure, and oxygen, not air, is used. Caking coals can be gasified without pretreatment. The coal is dried to between 1 and 8 percent moisture by weight and pulverized so that 70 to 90 percent will pass through a 200 mesh screen. The pulverized coal is conveyed with nitrogen from storage to the gasifier service bin and feed bins. Variable speed coal screw feeders then continuously discharge the coal into a mixing nozzle, where a mixture of steam and oxygen entrains the pulverized coal. Moderate temperature and high burner velocity prevent the oxidation of the coal.

The gasifier (Fig. 5-9) is a refractory-lined steel shell equipped with a steam jacket for producing low pressure process steam. A two-headed burner has heads 180° apart and can handle about 400 tons/day of coal. A four-headed gasifier, with burners 90° apart, can handle about 850 tons/day of coal. The burner heads are opposed to ensure that particles escaping from one burner will be burnt in the opposite burner. Gasifiers now being designed will operate at about 8 psig, and thus the exiting gases will have enough pressure to pass through venturi scrubbers. Carbon is oxidized in the gasifier producing a high temperature flame zone of about 3500°F. Heat losses and the endothermic carbon-steam reactions (Eq. 3.6) reduce the exit temperature to around 2700°F. The coal is almost instantaneously gasified and the latter is done nearly to completion. Carbon conversion, dependent on the reactivity of the coal, is about 96 to 98 percent. At the prevailing high operating temperatures, gaseous and vaporous hydrocarbons emanating from the coal decompose so rapidly that coagulation of coal particles during the plastic stage does not occur. Thus any coal can be gasified regardless of its caking property, ash content, or ash fusion temperature. Also, only gaseous products are produced; no tars, condensable hydrocarbons, or phenols are formed. Approximately 50 percent of the coal ash drops out as slag. The remainder of the ash leaves the gasifier as fine slag particles entrained in the exit gas.

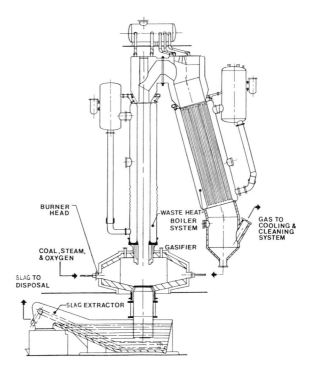

Figure 5-9
Koppers-Totzek two-headed gasifier.

Gas leaving the gasifier may be directly water quenched to solidify entrained slag droplets, if necessary. It then passes through a waste heat boiler where high pressure steam is produced. The gas exiting the waste heat boiler at 350° to 500°F is then scrubbed and cooled to approximately 95°F. The entrained solids are reduced to from 3×10^{-7} to 7×10^{-7} lb/scf. Particle-laden water from the gas scrubbing and cooling system is sent to a clarifier. The recovered clean water is cooled and recirculated through a gas washing system. The gas contains approximately 95 percent of the total sulfur in the coal, and this is removed by an appropriate process.

Table 5.14 presents a water equivalent hydrogen balance of the main process streams for a plant manufacturing 250×10^6 scf/day of medium-Btu power gas.[12] The coal is bituminous coal with a 16.5 percent moisture content and a heating value of 8,830 Btu/lb. The feed rate of as-received coal is 5,767 tons/day. For an output of 230×10^9 Btu/day, which is the approximate heating value of the

Table 5.14

Water equivalent hydrogen balance for Koppers-Totzek plant producing 250×10^6 scf/day of 303 Btu/scf power gas from 5,767 tons/day of as-received bituminous coal.*

	10^3 lb/hr	gal/10^6 Btu†
In		
Moisture in coal	79	3.0
Water equiv. of H_2 in coal	164	6.2
Steam to gasifier	72	2.7
Water to gasifier for spraying hot gas	42	1.6
Water to purification for makeup	2	0.1
Total	359	13.6
Out		
Water with wet slag from gasifier	4	0.2
Water with wet ash from scrubbing and cooling	54	2.0
Condensate from compression	30	1.1
Condensate in cooling (net)	21	0.8
Water vapor in vent gas from coal drier	82	3.1
Water equiv. of H_2 from purification (H_2S)	3	0.1
Water equiv. of H_2 in product gas	160	6.1
Water vapor in product gas	5	0.2
Total	359	13.6

*16.5 wt. % moisture, as-received heating value 8,830 Btu/lb.
†In the product gas.

standard size pipeline gas plant, the input water is approximately equivalent to $1,120 \times 10^3$ lb/hr of water.

5.4 Cooling Water

The unrecovered heat in any synthetic fuel plant has been defined as the heating value in the raw coal or shale that is not recovered in the synthetic fuel or byproducts. Part of the unrecovered heat is lost directly to the environment, while part is transferred indirectly through a heat transfer surface. How much heat is to be transferred indirectly, at what level, from what equipment, and from what sections of the plant must be known before the methods developed in Chapter 4 can be applied to define the appropriate cooling method. Once the cooling method is specified it is then a simple matter to determine the quantity of water evaporated.

Table 5.15
Sections of pipeline gas plants where heat is unrecovered
and sections where it is mostly recovered.

Unrecovered	Recovered
Gas Purification	Gasifier
Off-Gas Scrubbing and Quenching	Methanator
Steam Boilers	Shift Converter
Oxygen Production	Sulfur Plant
Electrical Generation and Use*	
Coal Drying	
Coal Feeding	
Product Cooling and Drying	

*Unrecovered electrical energy.

To define the disposition of unrecovered heat, the indirect cooling points, and the indirect cooling loads, it is necessary to carry out a heat balance around various sections of the plant and around the plant as a whole. Table 5.15 summarizes those sections in which heat is not recovered and those sections in which it is mostly recovered for plants designed for pipeline gas production from coal.[13]

To illustrate the details of a heat balance determination, we again choose the Synthane process. The plant design is the same as discussed in Section 5.2 and outlined in the flow diagram of Fig. 5-4. Since the plant must be in thermal balance, the energy in the feed coal not in the product gas or byproducts is the unrecovered heat. Table 5.16 presents a summary of the overall plant heat balance. An overall plant conversion efficiency of 65 to 70 percent may be considered representative for the production of pipeline gas by oxygen-blown

Table 5.16
Thermal balance for a 250×10^6 scf/day Synthane plant.

Heating Value	10^9 Btu/hr
Coal feed	17.1
Product gas	(9.8)
Tar	(0.8)
Char not used	(0.5)
Unrecovered heat	6.0
Overall conversion efficiency	65%

gasifiers. The exact value will always be specific to the process, design, coal, and climate.

Before examining the detailed breakdown of the plant heat loads, it should be emphasized that this methodology is not unique. It merely represents one accounting procedure to tally up the individual heat loads and to check the thermal balance of a given design. The actual method for cooling, on the other hand, is based on the principles laid out in Chapter 4.

First, the heat balance around the complete gasification train shown in Fig. 5-4 is considered. The balance is set down in Table 5.17. The enthalpies of liquid and gas are measured above those of liquid water and carbon dioxide, respectively, at 70°F. The enthalpies of the coal, char, and tar are the higher heating

Table 5.17
Heat balance around a 250 × 10⁶ scf/day Synthane gasification train.

Stream (Fig. 5.4)		Enthalpy (10⁹ Btu/hr)
	In	
①	Higher heating value of coal	17.08
②	Steam	1.13
⑦	Steam	0.30
	Total In	18.51
	Out	
⑬	Higher heating value of product gas	9.79
	Higher heating value of char	4.16
	Higher heating value of tar	0.80
	Steam produced	1.61
	Products and Byproducts Subtotal	16.36
⑤,⑨	Sensible heat of condensate	0.14
	Sensible heat of char	0.06
	Combustibles lost in gas purification	0.10
	Process dry cooling	1.33
	Process wet cooling	0.07
	Unaccounted losses	0.45
	Unrecovered Heat Subtotal	2.15
	Total Out	18.51

values calculated from their chemical compositions using the Dulong Formula (Eq. 3.1). The chemical compositions of the char and tar must be experimentally determined and verified by material and heat balances around the gasifier itself. The char leaving is quite hot, and therefore its sensible enthalpy should not be neglected. In the calculation of this enthalpy, a sufficiently accurate assumption is that the specific heat of the char is constant and equal to about 0.2 Btu/lb.

The enthalpy of the steam can be taken from any appropriate set of steam tables (for example, Ref. 14). The higher heating values of the more important gases considered are: CH_4 = 383,000 Btu/mole, H_2 = 123,000 Btu/mole, and CO = 122,000 Btu/mole. When calculating gas enthalpies, the specific heat variation with temperature[7,15] should be taken into account for the temperature ranges of these systems. Heat exchanger loads, whether wet cooled, dry cooled, or used to produce steam are calculated as the difference between inlet and outlet sensible heats of the noncondensable gases plus the difference in total enthalpy of any steam that may condense. In a reactor such as the methanator, short cuts are not possible and the total enthalpy of the inlet and outlet streams must be calculated.

Table 5.17 has been made to balance by inserting an entry for unaccounted losses. Note how small these losses are despite the simple procedure used to calculate each entry. For the most part the unaccounted losses are probably associated with radiation and convection from vessels and pipes. Another important point is that the unrecovered heat is quite small, representing less than 12 percent of the total input enthalpy. However, as shown in Table 5.16, the plant is not 88 percent efficient, so that it is necessary to consider where else the energy is consumed.

Some char is burned to provide energy to drive the plant. This may be seen most clearly by comparing the heating value of the char in Table 5.17 with the heating value left in the unused char listed in Table 5.16. Since most of the plant driving energy is not recovered, it is very important to discover where it goes. In fact, the heat balance around the gasification train is the least important prerequisite for determining cooling water requirements. The plant driving energy, as will be shown, is considerably more important.

The energy needed to drive the plant and its ultimate disposition is shown in Table 5.18. The energy required to separate oxygen from air is about 330 kwh/ton[16] and the energy to compress it to 1000 psia is about 120 kwh/ton; all of this energy is consumed by gas compressors. Another large compressor in the plant is needed for the lock hopper vessel into which the coal is loaded. In this hopper the coal, surrounded by carbon dioxide compressed to 1000 psia, is released into the

Table 5.18
Energy to drive a 250×10^6 scf/day Synthane plant and its ultimate disposition.

Energy Use and (Disposition)	10^9 Btu/hr
Oxygen production (70% turbines*, 30% compressors†)	1.27
Lock hopper compressors (70% turbines, 30% compressors)	0.08
Electricity generated—31,000 kw (70% turbines, 30% direct loss)	0.36
Net steam produced in process (100% used)	(0.18)
Acid gas removal (100% dry cooling)	1.01
Coal drying (100% direct loss)	0.42
Undiscovered requirements (100% wet cooling)	0.30
Stack loss—12% of char burnt (100% direct loss)	0.44
Total Char Used	3.70

*Turbine condensers.
†Interstage cooling.

gasifier. About 70 percent of the energy used for compressing gases ends up in steam turbine condensers and 30 percent in interstage coolers.

Another omission in doing the heat balance in Table 5.17 was that carbon dioxide was assumed to be separated in the acid gas removal system at no cost in energy. In fact, as shown in Table 3.5, regeneration of the acid gas absorbing liquor requires about 30,000 Btu/mole of CO_2 removed for the hot potassium carbonate system. This energy, listed in Table 5.18, is dissipated in the regenerator condenser, which is air cooled in this design.

A self-sufficient plant must produce its own electricity for the myriad of small uses required. In this design, two large uses are also included—the cooling water circulation pump and the acid gas absorbing liquor circulation pump. These two large pumps could be direct steam turbine driven. This would not affect the calculation, however, because the same turbine would be used to drive a generator or a pump, and electricity utilization is close enough to 100 percent efficient. About 70 percent of the energy used to produce electricity or to drive liquid pumps can be assumed lost in turbine condensers and 30 percent directly by radiation or convection.

As shown in Table 5.17, steam with a heat content of 1.43×10^9 Btu/hr is taken into the process, and 1.61×10^9 Btu/hr worth of steam is put out by the process. The steam put out is at a lower pressure and temperature than the steam taken in, but all of it can be used somewhere in the plant. The difference in heat content between the input and output of 0.18×10^9 Btu/hr is therefore subtracted in calculating the plant driving energy.

In the Synthane process, as for many of the synthetic fuel processes, the coal must be dried by heating before use. The energy to do this is proportional to the moisture vaporized and is all lost up the drier stack.

Although the major requirements for driving energy have been identified, water treatment and other minor requirements have not been evaluated. An arbitrary allowance of 10 percent is added to account for this. This allowance may be assumed to end up as a wet cooling load, to ensure a conservative estimate of the cooling water requirement. Much of the undetailed driving energy goes to separate phenol and ammonia from wastewater. To do this, as will be shown, wet cooled condensers are probably needed.

The net driving energy required is 3.26×10^9 Btu/hr, assumed to be supplied by burning char to generate steam. The boiler is assumed to be about 88 percent efficient, and about 12 percent of the heating value of the char that is burned will be lost up the stack.

From Table 5.18 it may be seen that the total energy taken in to drive the plant is 3.88×10^9 Btu/hr, or about 23 percent of the energy in the coal feed. This energy is consumed and represents unrecovered heat dissipated to the atmosphere. The energy is dissipated, 25 percent directly, 26 percent by dry cooling, 8 percent by wet cooling, 31 percent through steam turbine condensers (wet cooled or combined wet/dry cooled), and 10 percent through gas compressor interstage coolers (wet cooled or combined wet/dry cooled).

We have combined the results from Tables 5.17 and 5.18, summarizing in Table 5.19 the heat lost directly and by dry cooling, and in Table 5.20 the heat dissipated through wet cooling. The sum of all of these losses is 6.03×10^9 Btu/hr, which was the value listed in Table 5.16 corresponding to a plant efficiency of 65 percent. More importantly, from the point of view of water consumption, is that the results show under non-arid conditions that only 33 percent of the unrecovered heat goes to evaporating water, while 28 percent is lost directly to the atmosphere and 39 percent is dissipated by dry cooling. The total water evaporated in this case amounts to a little over 4×10^6 gal/day, where the quantity of water evaporated is estimated on 1,400 Btu transferred per pound of water evaporated (see Table 4.1). In the case of extreme aridity, where water consumption must be limited, wet/dry cooling of turbine condensers would be practiced, and as shown in Table 5.20 the water consumption could be cut by almost two thirds.

For the purposes of comparison, Table 5.21 shows the overall conversion efficiencies for the Hygas and Lurgi pipeline gas plants discussed in Section 5.3. As expected, the Hygas process shows a somewhat higher conversion efficiency

Table 5.19
Disposition of unrecovered heat lost directly and by
dry cooling in a 250 × 10⁶ scf/day Synthane plant.

	10^9 Btu/hr
Direct Loss	
Coal drying	0.42
Stack loss	0.44
Electricity generated	0.11
Sensible heat of condensate and char	0.20
Combustibles lost in gas purification	0.10
Unaccounted losses in gasification train	0.45
Total Direct Loss	1.72
Dry Cooling	
Acid gas removal	1.01
Process dry cooling	1.33
Total Dry Cooling	2.34

because of the higher methane production in the gasifier. A point which must be emphasized, however, is that the conversion efficiency using any given process may be expected to range over several percentage points, depending upon the details of the individual plant design, the coal, the climate, and other factors. For example, the Lurgi plant conversion efficiency shown in Table 5.21 is based upon the El Paso design using a New Mexico coal.[9-11]. A design for North

Table 5.20
Disposition of unrecovered heat lost by wet cooling and water
evaporated in a 250 × 10⁶ scf/day Synthane plant.

	Heat Lost	Water Evaporated	
Load	10^9 Btu/hr	10^3 lb/hr	gal/10⁶ Btu†
Turbine condensers*	1.20	857	10.5
Compressor interstage cooling*	0.40	286	3.5
Process and driving energy wet cooling	0.37	264	3.2
Total Wet Cooling	1.97	1,407	17.2

*In arid regions or where water is expensive, 10 percent of cooling load on turbine condensers and 50 percent of load on interstage coolers is assumed wet cooled. Total evaporated water becomes 490 × 10³ lb/hr or 6.0 gal/10⁶ Btu.
†In the product gas.

Table 5.21
Thermal balance for 250×10^6 scf/day Lurgi and
Hygas plants.

	10^9 Btu/hr	
Heating Value	Lurgi	Hygas
Coal feed	14.6	14.2
Product gas	(9.9)	(10.1)
Unrecovered heat	4.7	4.1
Overall conversion efficiency	68%	71%

Dakota with the same output but using lignite shows a conversion efficiency of 65 percent.[17]

An important concern here is the quantity of unrecovered heat that goes to evaporating water. A detailed thermal balance similar to that presented for the Synthane plant has been carried out for the Hygas design outlined and discussed in Section 5.3.[18] The results of this balance are shown in Tables 5.22 and 5.23. The sum of all unrecovered losses amounts to 4.08×10^9 Btu/hr, but only 26 percent of this heat goes to evaporating water, while 41 percent is lost directly to the environment and 33 percent is dissipated by dry cooling. For the Lurgi example,[9–11] $1,217 \times 10^3$ lb/hr of water is evaporated for cooling, indicating that

Table 5.22
Disposition of unrecovered heat lost directly and by dry
cooling in a 250×10^6 scf/day Hygas plant.

	10^9 Btu/hr
Direct Loss	
Coal drying	0.26
Stack loss	0.20
Electricity generated, slurry pumps	0.10
Sensible heat of condensate	0.01
Combustibles lost in gas purification	0.80
Sensible heat and heating value of ash	0.32
Total Direct Loss	1.69
Dry Cooling	
Acid gas removal	0.80
Process dry cooling	0.55
Total Dry Cooling	1.35

Table 5.23
Disposition of unrecovered heat lost by wet cooling and water
evaporated in a 250 × 10⁶ scf/day Hygas plant.

Load	Heat Lost 10^9 Btu/hr	Water Evaporated 10^3 lb/hr	gal/10^6 Btu†
Turbine condensers*	0.65	464	5.52
Compressor interstage cooling*	0.16	114	1.36
Process and driving energy wet cooling	0.23	164	1.95
Total Wet Cooling	1.04	742	8.83

*In arid regions or where water is expensive, 10 percent of cooling load on turbine condensers and 50 percent of load on interstage coolers is assumed wet cooled. Total evaporated water becomes 268 × 10³ lb/hr or 3.2 gal/10⁶ Btu.
†In the product gas.

about 36 percent of the unrecovered heat is dissipated by evaporative cooling. However, in a different design at the same rate, with the same coal, output, gasifier, and conversion efficiency, only about 26 percent of the unrecovered heat is dissipated by wet cooling.[19,20] This emphasizes again that the differences in consumptive water use are in large measure a design choice and not one of fundamental need. No two plants should be expected to use exactly the same amounts of water.

5.5 Quality of Effluent Process Water Streams

We have so far concentrated on defining the quantities of the various water streams in coal gasification, with little attention to quality. If effluent water streams from the process are to be recycled or cleaned up and reused, then it is necessary to define their quality, as measured by the nature and extent of the contaminants. In general, four broad groups of contaminants may be defined: physical, chemical, biological, and physiological. In water supply for domestic consumption the last two groups are very important. However, physiological factors such as taste and odor essentially play no role in our considerations, while biological factors are in large part limited to concerns of bacterial and algae buildup that can lead to fouling of heat transfer surfaces in cooling water systems; this will be examined in Chapter 8. In dealing with physical contamination, the most important example is solid or liquid suspended matter such as ash or char particles, tars, and oils.

In terms of effluent water treatment in synthetic fuel plants, chemical contaminants are the most numerous and generally the most difficult to remove. The chemical pollutants may be either inorganic or organic. Among the most important categories of inorganic pollutants are soluble gases, acids and bases, hardness, heavy metals, and soluble salts. The foul process condensates from all gasification reactors, except those operating at high temperature, contain large amounts of dissolved ammonia formed from nitrogen in the coal. Ammonia is usually present in such high concentrations that releasing it to the atmosphere, as in a cooling tower, would give a highly noxious odor. Ammonia can be corrosive to copper piping and inhibit biological decomposition of the organic contaminants. The gasifier condensates also contain large quantities of the dissolved acid gas carbon dioxide and smaller quantities of hydrogen sulfide. Carbon dioxide, a strong corrosive, is mostly hydrolyzed in solution to carbonic acid, which ionizes to varying concentrations of bicarbonate (HCO_3^-) and carbonate ($CO_3^=$) ions, depending on the hydrogen ion concentration. When a large excess of ammonia is present, as is often the case, the dissolved carbon dioxide is over 96 percent bicarbonate. Hydrogen sulfide is present unionized and as the hydrosulfide ion (HS^-). In hard alkaline waters high in calcium, the scale forming compound calcium carbonate will precipitate, which can seriously impair the effectiveness of heat transfer surfaces. Heavy metals in the raw coal may, for example, be scrubbed out of the flue gases. Chlorine present in the raw coal will also appear as chloride in the process condensates.

Organic contaminants are present in all foul process condensates in large quantities. Among the most prominent are phenols (carbolic acid), cresols, and "fatty" organic acids (carboxylic acids). Some of the organic matter may be removed by biological decomposition. When the decomposition takes place in the presence of free oxygen, carbon dioxide and water are produced, together with a settleable bacterial sludge. In this way, part of the organic matter is oxidized and part is converted to cell material. This type of breakdown, caused by aerobic bacteria, is termed aerobic decomposition. If the breakdown occurs in the absence of free oxygen and the bacteria use combined oxygen, then carbon dioxide, methane, and some inert material are the end products. This is termed anaerobic decomposition. Many of the organic compounds found in the effluent process streams cannot or can only very slowly biologically decompose. These organics are "refractory" in that they resist biological decomposition. Under suitable conditions, however, they can be chemically oxidized to carbon dioxide and water. Phenols, which can be easily biodegraded when dilute, are toxic in high concentrations and inhibit biological oxidation.

There are a number of accepted parameters by which water quality is defined. Among the more important nonspecific parameters are total dissolved solids (TDS), salinity, pH, alkalinity, and hardness. TDS is simply the residue in filtered water remaining after evaporation. Drinking water standards generally specify less than 500 milligrams per liter (mg/l), 1 mg/l being 1 part per million by weight (ppm). Salinity is measured by the amount of chloride in the water and is about 1.8 times the chloride content. For reference, drinking water standards specify less than 250 mg/l of chloride, while seawater contains about 20,000 mg/l.

The pH of the water is defined by $pH = -\log[H^+]$, where $[H^+]$ is the hydrogen ion concentration in moles/liter. For pure water $[H^+] = 10^{-7}$ moles/liter so that the neutral pH is 7. When an acid is added to water the hydrogen ion concentration increases resulting in a lower pH. When a base is added, the hydroxyl ions combine with the free hydrogen ions, lowering the hydrogen ion concentration and raising the pH. The pH scale from 0 to 7 is the acid range and from 7 to 14 the base range. Alkalinity is a measure of the ability to absorb a strong mineral acid with a limited decrease in pH. This ability can be given by the hydroxyl ion, that is, a water with a high pH has a high alkalinity. Alkalinity is predominantly a measure of dissolved carbon dioxide in all forms. When a strong mineral acid is added, carbonate and bicarbonate are forced to unionized carbonic acid, and carbon dioxide is released from the solution. Many waters have a high alkalinity caused by a high content of carbon dioxide, even though the pH is close to neutral.

Hardness is simply the concentration of metallic ions, principally calcium and magnesium, that can precipitate soap. For instance, soft water is considered to be less than 60 mg/l of hardness and very hard water greater than 180 mg/l.

Among the most common parameters for the measurement of organic content in a wastewater are the biochemical oxygen demand (BOD), chemical oxygen demand (COD), and total organic carbon (TOC). By definition, BOD is the amount of dissolved oxygen used by microorganisms in the aerobic oxidation of organic matter in a sample of wastewater at 20°C. The BOD of wastewater is time dependent because the carbonaceous oxygen demand progresses at a decreasing rate with time, since the rate of biological activity decreases as the available food supply decreases. The most frequently used period for the BOD parameter is 5 days (BOD_5), and it is usually reported in mg/l. For reference, a strong municipal wastewater may have a BOD_5 of 250 mg/l, while the value for a phenolic process condensate might be 20,000 mg/l. The chemical oxygen demand (COD) is the amount of oxygen required for chemical oxidation of the organic matter to carbon dioxide and water by strong oxidants under acid conditions. No uniform relationship exists between COD and BOD, except that COD must always be greater than BOD to the extent that refractory or nonbiodegradable organic matter is represented in the total organic content. Process condensates can have COD-to-BOD ratios as high as 1.5-to-2. The total organic carbon analysis (TOC) involves oxidizing the organic carbon at a temperature of about 950°C in the presence of a catalyst in a stream of oxygen. The amount of carbon dioxide that results is measured, and this provides a determination of the carbon content in the organic matter.

In coal gasification plants, there are four key process water effluent streams for which water quality analyses are needed if recycling or treatment for reuse, such as for boiler feed, is to be done. The dirtiest water stream is the foul process condensate recovered after gasification. Streams of intermediate cleanliness are those recovered after shift conversion and gas purification. The water of reaction stream from methanation is quite clean. Not every plant has all of the streams. In power gas production the shift conversion and methanation streams are absent. Some streams are also absent in some pipeline gas plants as, for example, the shift conversion and gas purification condensates in the Hygas process. Moreover, the term ''dirty'' must be qualified as referring to the low and intermediate temperature gasifiers and not to the very high temperature gasifiers such as Koppers-Totzek, where the condensate from the gasifier is relatively clean (see Tables 5.1–5.3).

The quality of the foul process condensate is directly linked to the gasifier process, defined by its operating variables, and to the rank of coal. In Table 5.1 a very broad measure of the off-gas cleanliness is indicated for different gasifier types based on operating temperature, the highest temperature reactors developing the cleanest gas product. Other process variables that affect the effluent quality are: gas residence time, heatup rate of the coal, and gas-solid mixing. Massey and others[21] have studied these variables in a Synthane gasifier and showed that increasing the reaction temperature and the coal heatup rate reduces the COD, TOC, and tar production without any significant alteration of the off-gas characteristics. They also showed that increasing the gas residence time (and overall gas-solid contacting) at an approximately constant coal heatup rate and reaction gas temperature results in a decrease in phenol production and a somewhat less pronounced decrease in COD, TOC, and tar production, again without changing the product gas.

On the basis of these results, the off-gas condensate characteristics for a given coal may be roughly deduced for the processes discussed in Sections 5.2 and 5.3. The Lurgi (non-slagging) and Synthane gasifiers should give the dirtiest water.

The Hygas reactor, because of its higher temperature at the top and faster coal heatup rate, might be expected to yield a somewhat cleaner condensate water with less organic contamination. The CO_2 Acceptor, which has a fluid bed of long residence time, should give a relatively clean condensate. The condensate from the high temperature Koppers-Totzek gasifier should be quite clean. These generalizations are substantiated by the analyses presented below. A limited assessment of the Hygas and CO_2 Acceptor waters by Massey[22] supports the conclusions about these gasifiers.

Analyses of foul process condensate from the Synthane gasification of two different rank coals are shown in Table 5.24.[23] Clearly, the organic contamination, which is mostly dissolved, is significantly affected by coal rank. For both coals the contamination is high. Phenol appears as a major component of the contamination; the chemical oxygen demand (COD), however, represents other organic contaminants in addition to phenol. To convert phenol to carbon dioxide and water requires 7 moles of oxygen for each mole of phenol, which is equivalent to 2.38 mg of oxygen per mg of phenol. This is termed the theoretical oxygen demand (ThOD), and for the phenol concentrations listed, constitutes about 40 percent of the COD for both coals. Analyses of Lurgi process conden-

Table 5.24
Analysis of foul process condensate from Synthane
gasification of two coals, mg/l (except pH).

	Illinois #6 C Bituminous	North Dakota Lignite
pH	8.6	9.2
Suspended Solids	600	64
COD	15,000	38,000
TOC	4,300[a]	11,000[b]
Inorganic Carbon	3,400[c]	6,500[b]
Phenol as C_6H_5OH	2,600	6,600
Ammonia	8,100	7,200
Thiocyanate	152	22
Chloride	500	—
Total Sulfur	1,400	—

[a]Estimated from COD/TOC = 3.5.
[b]Scaled from Ref. 21.
[c]From CO_3 and HCO_3 analyses.

Table 5.25
Analyses of foul process condensate from Hygas
gasification of two coals, mg/l.

	Illinois #6 C Bituminous	Montana Lignite
COD	3,000	13,600
BOD$_5$	2,680	—
TOC	700	3,900
Inorganic Carbon	520	2,500
Phenol as C$_6$H$_5$OH	270	1,200[a]
Ammonia	8,700	3,400
Chloride	260	—

[a]Estimated at 21% of COD and 2.38 mg COD/mg phenol.

sate[24] show comparable total organic contamination, although the phenol constitutes from 45 to 90 percent of the COD.

Table 5.25 gives analyses of the foul process condensate from the Hygas gasification of two different rank coals.[18] These represent single measurements but they are within the ranges reported by Luthy and others.[25] The measured phenol for the Illinois coal represents about 20 percent of the COD, and this fact and the theoretical oxygen demand were used to provide an estimate of the phenol produced with the Montana lignite. From a comparison of Tables 5.24 and 5.25, the Hygas condensate has less organic contamination than the Synthane condensate. In broad terms the Hygas process generates a process condensate of intermediate quality.

The CO$_2$ Acceptor and Koppers-Totzek process condensates are quite clean waters. Massey[22] and Luthy and others[25] have reported negligible organic contamination in the CO$_2$ Acceptor condensate. The cleanliness of the water from the high temperature Koppers-Totzek process is illustrated in Table 5.26 by the analysis of this water after using it to quench the slag from the gasifier.[26]

Ammonia is a major contaminant in all the waters. It is derived from the nitrogen in the coal, although only a fraction of this coal nitrogen (about 10 to 25 percent) ends up as ammonia in the condensate. At high temperatures, as in the Koppers-Totzek process, ammonia is not formed and instead free nitrogen is released. The sulfur in the coal is mostly converted to hydrogen sulfide. The carbon dioxide, present in large amounts in the condensate, is a stronger acid than hydrogen sulfide, which is present in the condensate in only small amounts.

Chloride is a significant constituent if the condensate is treated to the quality of

Table 5.26
Analysis of process condensate from
Koppers-Totzek coal gasification,
Kutahya, Turkey, after use as slag quench,
mg/l (except pH).

pH	8.9
COD	63
Ammonium	122
Chloride	46
Total Sulfur	58
Hardness	227

boiler feed. Chlorine in the coal is converted to hydrochloric acid by the conversion process, and the only important way it leaves the plant is in the condensate. The chloride concentration in the water, however, will depend as much on the condensate rate (which depends on the process conditions and whether wet or dry coal is used) as it does on the chlorine content of the coal. Wide variations in chloride concentration in the condensate may therefore be expected among different coals and designs.

There may be three additional key process effluent streams in a coal gasification plant. The streams of intermediate cleanliness with respect to the process condensate are those recovered after shift conversion and after gas purification. Analyses of these streams are not available but, if both are present, we may assume that they are mixed and that they yield a medium quality condensate. Since most of the condensable and water soluble contaminants in the gasifier off-gas appear in the foul condensate, the mixed condensate may be considered typically to have concentrations one tenth of those for the corresponding foul condensate.

In a pipeline gas plant the cleanest condensate is obtained from the water of reaction in methanation. This water is very clean because it comes from clean conditions. A sample of this methanation water is crystal clear and without odor. The condensate is saturated with CO_2, CO, CH_4, and H_2, but it is always recovered hot and under pressure, and therefore these gases are assumed to be removed by flashing. After flashing the condensate can be assumed to have less than 200 mg/l of total dissolved solids and to be suitable as a feed to boiler water deionization treatment.

We conclude this section by observing that there are a number of systematic relationships between the effluent characteristics and the process and coal but that any given condensate quality must necessarily be specific to the particular design and particular coal.

5.6 A Summary of Water Stream Quantities

In Table 5.27 all of the process stream water quantities found for the plants designed to produce 250×10^6 scf/day of pipeline gas are assembled in summary form. These streams, shown in Fig. 5-1, do not include water for quenching ash or slag, since this water requirement depends more on the ash content of the coal than on the process. Data on water streams for the Bi-Gas and Kellogg Molten Salt processes from Ref. 27 (see also, Ref. 18) have been included in Table 5.27, although these processes were not discussed in any detail in the text. This is done in order to present a representative picture of the pipeline gas water streams for each of the gasifier types classified in Section 5.1. It must be emphasized that all the results in Table 5.27 are individual examples and are not the only reasonable set of numbers.

The total fresh water requirement for all of the plants has a relatively narrow range, from the low value of about 10^6 lb/hr for the CO_2 Acceptor and Hygas processes to about 1.2 to 1.8×10^6 lb/hr for the other processes. The higher temperature gasifiers have a somewhat higher requirement.

The only process discussed that accepts wet coal is the Lurgi process. However, the coal moisture is not consumed in the reaction but appears as dirty condensate, which must be cleaned; hence the high value for the Lurgi dirty

Table 5.27
Summary of process water streams in 10^3 lb/hr for the production of 250×10^6 scf/day of pipeline gas.

		In		Out	
Type	Process	Coal Moisture	Fresh Water	Dirty Condensate	Methanation Water
Entrained Flow	Bi-Gas	12	1,776	996*	242
Fixed Bed	Lurgi	273	1,635	1,085	274
Fluid Bed	Synthane	69	1,167	674	140
Hydrogasifier	Hygas	21	1,015	294	201
Indirect Heat	CO_2 Acceptor	—	1,035	270*	256
Molten Media	Molten Salt	22	1,507	745	313

*These condensates are relatively clean.

condensate stream. Some of the dirty water leaving the different processes depends on the gas purification method and its operation. The lowest effluent rates of about 0.3×10^6 lb/hr are for the CO_2 Acceptor and Hygas processes. These are the processes with the lowest fresh water requirement. All the other processes put out between 0.7×10^6 and 1.1×10^6 lb/hr of dirty water.

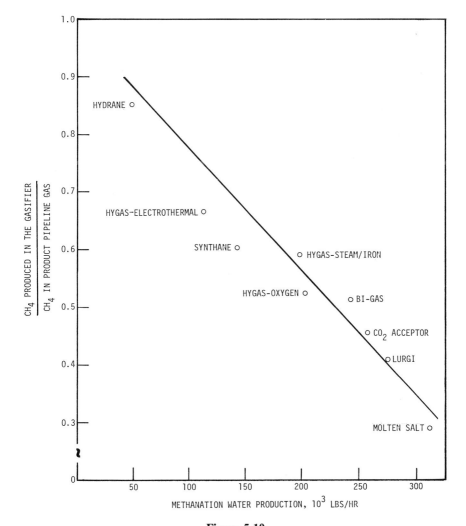

Figure 5-10
Quantity of methanation water as a function of methane produced in gasifier for 250×10^6 scf/day pipeline gas plants.

The very clean water formed in the methanation reaction can be treated to high purity boiler feed water quite easily. The actual quantity of water evolved depends on the relative output of methane in the gasifier and the subsequent degree of methanation required. This is illustrated graphically in Fig. 5-10.[13] Data are plotted there for all of the processes included in Table 5.27 and a few additional ones not discussed previously.

In Table 5.28 the net water requirements for the different processes are shown. These requirements are the differences between the total imported and exported streams. The net requirements range from about one half to one third of the imported fresh water, providing some indication of the result of reuse. The net requirements are also shown normalized in gal/10^6 Btu in the product gas. Quite interestingly, the values are very close to 6 gal/10^6 Btu for all the processes examined.

Because less hydrogen is required for power gas than for pipeline gas, the process water requirements should be less for low- and medium-Btu gas production than for pipeline gas production. That this is indeed the case is illustrated by the Koppers-Totzek power gas example summarized in Table 5.14. The total imported water stream for that case is about 7.4 gal/10^6 Btu or about half that for pipeline gas production. On the other hand, the difference between the imported and exported streams is about 3.3 gal/10^6 Btu, or less than half that of the imported stream, again indicating the important savings that can result from reuse. These general results for power gas production are supported by calculations for other processes.[27]

The results of this chapter have shown that the overall conversion efficiency of

Table 5.28
Difference between imported and exported process water streams
for the production of 250 × 10^6 scf/day of pipeline gas.

Process	Net Water Requirement		
	10^3 lb/hr	10^6 gal/day	gal/10^6 Btu*
Bi-Gas	550	1.58	6.72
Lurgi	549	1.58	6.63
Synthane	422	1.22	5.17
Hygas	541	1.56	6.43
CO_2 Acceptor	509	1.47	6.16
Molten Salt	471	1.36	5.94

*In the product gas.

pipeline gas plants is in the range of 65 to 70 percent. The unrecovered heat that is dissipated to the environment is therefore about 1×10^{11} to 1.3×10^{11} Btu/day. In general, no more than 25 to 35 percent of this heat goes to evaporating water. Assuming 1,400 Btu transferred per pound of water evaporated, this gives a range of 2×10^6 to 4×10^6 gal/day of water evaporated in the wet cooling of these plants. This is equivalent to 700×10^3 to $1,400 \times 10^3$ lb/hr, which is more than the dirty condensate usually exported from the process. The usual use of dirty condensate is for cooling water. In arid regions where water is scarce, combined wet/dry cooling of the steam turbine condensers could cut this requirement by two thirds. The makeup water, as pointed out in Section 4.6, may be estimated at 10 percent of the evaporated water, with the extra being blown down. The blowdown does not, however, necessarily represent a consumption so long as it is reused within the plant.

Chapter 5. References

1. Dravo Corp., "Handbook of Gasifiers and Gas Treatment Systems," Report No. FE-1772-11, Energy Res. & Develop. Admin., Washington, D.C., Feb. 1976.

2. Hendrickson, T. A., *Synthetic Fuels Data Handbook*. Cameron Engineers, Inc., Denver, Colo., 1975.

3. Schnider, L. A. and Fischer, D. D., "In Situ Coal Gasification a Unique Means of Energy Recovery," *Mechanical Engineering* **98** (3), 18–24, March 1976.

4. Nadkarni, R. M., Bliss, C., and Watson, W. I., "Underground Gasification of Coal," *Clean Fuels from Coal Symposium II,* pp. 625–651, Institute of Gas Technology, Chicago, Ill., 1975.

5. Hebden, D., "High Pressure Gasification under Slagging Conditions," *Proc. Seventh Synthetic Pipeline Gas Symposium,* pp. 385–400, American Gas Association, Arlington, Va., 1975.

6. U.S. Dept. of the Interior, "SYNTHANE Gasification at 1,000 psia, Followed by Shift Conversion, Purification, Single-Stage Tube Wall Methanation and Pollution Control. 250 Million-SCFD High-Btu Gas Plant. Wyodak Seam Coal. An Economic Analysis," Report No. 75-15, Bureau of Mines, Morgantown Energy Research Center, Morgantown, W.Va., 1974.

7. Girdler Chemical Inc., *Physical and Thermodynamic Properties of Elements and Compounds.* Girdler Chemical, Inc., Louisville, Kentucky.

8. Clancey, V. T., Marwig, U. D., and Lindahl, D. R., "Pipeline Gas from Lignite Gasification. Current Commercial Economics," OCR R&D Report No. 16-Interim Report, Office of Coal Research, Washington, D.C., 1969.

9. El Paso Natural Gas Co., "Second Supplement to Application of El Paso Natural Gas Company for a Certificate of Public Convenience and Necessity," Federal Power Commission Docket No. CP73-131, 1973.

10. Gibson, D. R., Hammons, G. A., and Cameron, D. S., "Environmental Aspects of El Paso's Burnham I Coal Gasification Complex," *Symp. Proc.: Environmental Aspects of Fuel Conversion Technology,* pp. 91–100, Report No. EPA-650/2-74-118, Environmental Protection Agency, Research Triangle Park, N.C., Oct. 1974.

11. Milios, P., "Water Reuse at a Coal Gasification Plant," *Chemical Engineering Progress* **71** (6), 99–104, June 1975.

12. Magee, E. M., Jahnig, C. E., and Shaw, H., "Evaluation of Pollution Control in Fossil Fuel Conversion Processes. Gasification, Section 1: KOPPERS-TOTZEK Process," Report No. EPA-650/2-74-009a, Environmental Protection Agency, Research Triangle Park, N.C., Jan. 1974.

13. Goldstein, D. J. and Probstein, R. F., "Water Requirements for an Integrated SNG Plant and Mine Operation," *Symp. Proc.: Environmental Aspects of Fuel Conversion Technology II,* pp. 307–332, Report No. EPA-600/2-76-149, Environmental Protection Agency, Research Triangle Park, N.C., June 1976.

14. Keenan, J. H., Keyes, F. G., Hill, P. G., and Moore, J. G., *Steam Tables.* Wiley, New York, 1964.

15. Keenan, J. H. and Kaye, J., *Gas Tables.* Wiley, New York, 1957.

16. Hugill, J. T., "Cost Factors in Oxygen Production," *Symposium on Efficient Use of Fuels in Metallurgical Industries,* pp. 151–165, Institute of Gas Technology, Chicago, Ill., Dec. 1974.

17. Michigan-Wisconsin Pipeline Co. and American Natural Gas Coal Gasification Co., "Application for Certificate of Public Convenience and Necessity Before the Federal Power Commission," Docket No. CP75-27B, 1975.

18. Goldstein, D. J. and Yung, D., "Water Conservation and Pollution Control in Coal Conversion Processes," Report No. EPA-600/7-77-065, Environmental Protection Agency, Research Triangle Park, N.C., June 1977.

19. Western Gasification Co., "Amended Application for Certificate of Public Convenience and Necessity," Federal Power Commission Docket No. CP73-211, 1973.

20. Berty, T. E. and Moe, J. M., "Environmental Aspects of the Wesco Coal Gasification Plant," *Symp. Proc.: Environmental Aspects of Fuel Conversion Technology,* pp. 101–106, Report No. EPA-650-2/74-118, Environmental Protection Agency, Research Triangle Park, N.C., Jan. 1974.

21. Massey, M. J., Nakles, D. V., Forney, A. J., and Haynes, W. P., "Role of Gasifier Process Variables in Effluent and Product Gas Production in the Synthane Process," *Symp. Proc.: Environmental Aspects of Fuel Conversion Technology II,* pp. 153–177, Report No. EPA-600/2-76-149, Environmental Protection Agency, Research Triangle Park, N.C., June 1976.

22. Massey, M. J., "An Approach to Residuals Management in Coal Conversion Processing," *Proc. First Symp. on Management of Residues from Synthetic Fuel Production* (J. J. Schmidt-Collerus and F. S. Bonomo, eds.), pp. 31–54, University of Denver, Denver Research Institute, Denver, Colo., 1976.

23. Forney, A.J., Haynes, W. P., Gasior, S. J., Johnson, G. E., and Strakey, J. P., Jr., "Analyses of Tars, Chars, Gases and Water Found in Effluents from the Synthane

Process," Technical Progress Report No. 76, Bureau of Mines, Dept. of the Interior, Pittsburgh Energy Research Center, Pittsburgh, Penn., Jan. 1974.

24. Woodall-Duckham Ltd., "Trials of American Coals in a Lurgi Gasifier at Westfield, Scotland," Report No. 105 (NTIS Catalog No. FE-105), Energy Res. & Develop. Admin., Washington, D.C., Nov. 1974.

25. Luthy, R. G., Massey, M. J., and Dunlap, R. W., "Analysis of Wastewater from High Btu Coal Gasification Plants," *Proc. 32nd Industrial Waste Conference, May 10–12, 1977,* Purdue Univ., Lafayette, Indiana, 1977.

26. Farnsworth, J. F., Mitsak, D. M., and Kamody, J. F., "Clean Environment with Koppers-Totzek Process," *Symp. Proc.: Environmental Aspects of Fuel Conversion Technology,* pp. 115–130, Report No. EPA-650/2-74-118, Environmental Protection Agency, Research Triangle Park, N.C., Jan. 1974.

27. Probstein, R. F., Goldstein, D. J., Gold, H., and Shen, J. S., "Water Needs for Fuel-to-Fuel Conversion Processes," *Water-1975, AIChE Symp. Series,* Vol. 71, No. 151, pp. 209–220, 1975.

6
Liquid and Solid Fuel Production

6.1 Liquefaction and Clean Coal Technologies

In Chapter 3 pyrolysis and hydrogenation were shown to be the principal routes by which clean liquid and solid fuels could be derived from coal and oil shale. In the following discussion all of the hydrogenation technologies will be for coal conversion and pyrolysis technologies will be for either coal or shale as the raw fuel. Pyrolysis, however, may take place in the presence of hydrogen, although in any case the oils produced by the pyrolysis of coal or shale will be upgraded by refinery-type hydrotreating procedures.

In discussing the technologies to produce clean liquid or solid fuels, the kind of process—direct hydrogenation, indirect hydrogenation, or synthesis from medium-Btu gas—must first be distinguished.

Hydroliquefaction is the direct hydrogenation of coal at high pressure and the principal direct procedure used. In this method pulverized and dried coal is slurried with recycled oil, mixed with hydrogen, and fed to a catalytic reactor operated at a moderate temperature (about 850°F) and high pressure (2,000 to 4,000 psig). The reactor may use entrained, fluid bed, or fixed bed catalysts. The H-Coal process uses a fluidized bed reactor operating at 2,250 to 3,000 psig in which the upward passage of slurry and gases maintain the catalyst in a bubbling, fluidized state. The principal product is a synthetic crude or a low sulfur fuel oil and a high-Btu gas. In the Synthoil process, the slurry of coal and oil is fed to a fixed bed catalytic reactor operating at 2,000 to 4,000 psig and propelled through the reactor by the hydrogen gas that is mixed in. The principal product is a low sulfur fuel oil. These and other processes are cited in Ref. 1. In hydroliquefaction, most of the oxygen in the coal is converted to water and the remainder to carbon dioxide. The sulfur and nitrogen are converted to hydrogen sulfide and ammonia, respectively.

Solvent Extraction is an indirect method of hydrogenation in which coal is partially dissolved in a coal-derived solvent that exchanges hydrogen with the coal. Pulverized and dried coal is slurried with recycled solvent. In the Solvent Refined Coal (SRC) process, hydrogen is mixed with the slurry prior to the dissolution of the coal in a reactor at a temperature of 800° to 900°F and a pressure of 1,000 to 2,000 psig. The hydrogen reacts with the solvent and the solvent then gives its hydrogen to the coal. When the coal dissolves in the reactor, the ash and pyritic sulfur can be separated out. When this is the case, most of the hydrogen is consumed in forming water, hydrogen sulfide, and ammonia (see Eqs. 3.10–3.12), and there is only a relatively low degree of hydrogenation. Alternatively, the recycled solvent can be hydrogen-rich from a separate catalytic hydrogenation, as in the Exxon Donor Solvent Process (EDS),[2] where the dissolution takes place at a temperature of 700° to 900°F and a pressure of 1,500 to 2,500 psig. A much greater degree of hydrogenation results. The solvent is recovered from the higher boiling point product by distillation.

In the SRC process the undissolved solids are filtered out prior to distillation, the solvent is recycled directly, and the heavy product is cooled and solidified to produce a pitch-type solvent refined coal. In the EDS process the undissolved solids are removed with the distillate bottoms, the solvent is recycled to the reactor after catalytic hydrogenation, and the principal products are raw liquid fuels (naptha and low sulfur fuel oil). As in direct hydrogenation, most of the oxygen in the dissolved coal is converted to water.

Liquid Hydrocarbon Synthesis or Fischer-Tropsch synthesis starts with a medium-Btu gas that has been "shifted" and purified to produce synthesis gas that is composed principally of carbon monoxide and hydrogen. As described in Section 3.5, the synthesis gas is reacted in the presence of a catalyst to produce a variety of liquid hydrocarbons represented by the reactions (3.13) and (3.14). In the commercial scale Sasol plant in South Africa, a fixed bed catalytic reactor is operated in parallel with a fluid bed reactor. The operating conditions for the fixed bed are 430° to 490°F and 360 psig and for the fluid bed they are 600° to 625°F and 330 psig.[3] A larger plant now under construction by Sasol to produce 50,000 barrels/day of gasoline incorporates only the fluid bed reactor. With different catalysts than those used for Fischer-Tropsch synthesis, oxyhydrocarbons such as methanol can be produced (see Eq. 3.15). Typical operating conditions for the synthesis of methanol and its coproducts are in the range of 1,500 to 5,000 psig and 575° to 750°F. Methanol, however, is also manufactured from synthesis gas at the relatively low pressure of 800 psig at 500°F using an active copper, zinc, chromium catalyst.[1] The higher temperatures and pressures yield larger fractions of other materials such as ethers and higher alcohols. It is important to emphasize here that synthesis technology starting with coal is inherently inefficient and hence has a high cooling load, since it involves first producing a synthesis gas from the coal and then forming the gas into a liquid.

Pyrolysis, as described in some detail in Section 3.3, yields condensable tar, oil, and water vapor, and noncondensable gases through the destructive distillation of coal or oil shale. In the case of coal, char is rejected. The condensed pyrolysis product must, however, be further hydrogenated to remove the sulphur and nitrogen and to improve the liquid fuel quality. Essentially, all of the conversion processes based on pyrolysis are similar. They differ principally in their methods and rates of applying the heat, their ultimate temperatures, and their gas atmospheres. For purposes of illustration, several potentially commercial processes are cited here. A listing and classification of the many pyrolysis processes, particularly those for oil shale, are reserved for the sections that follow on pyrolysis technologies.

In the Garrett coal pyrolysis process,[5] hot char from a separate entrained bed furnace provides the heat for the flash pyrolysis of fine coal in a reactor at a temperature of about 1100°F and approximately atmospheric pressure. Pilot plant yields show a little less than 60 percent by weight of char and 35 percent of tar, and the remainder gas (700 Btu/scf) and water.

A fluidized bed coal pyrolysis process, which is carried out in successive stages at successively higher temperatures, is the COED process.[6] By partially devolatilizing the coal at a lower temperature, the coal can be heated to a higher temperature in the next reactor without agglomerating and plugging the fluid bed. The process is designed for four stages with temperatures dependent on the coal caking properties, ranging from about 550°F in the first stage to about 1550°F in the last stage and with pressures from about 5 to 10 psig. Heat for the process is generated by burning char in the last stage with a steam-oxygen mixture and then using the hot gases and the hot char from this stage to heat the other vessels. The process yields char, pyrolysis oils, and a medium-Btu gas. Typical pilot plant yields show the char accounts for more than 60 percent by weight of the total of the three products and the oil about 20 percent.[6]

Two processes for the pyrolysis of either coal or oil shale that use the same reactor and the same flow sheet but different names are the TOSCOAL process for coal and the TOSCO II process for oil shale. In this scheme crushed coal or shale is fed to a horizontal rotating kiln where the material is heated by hot ceramic balls to the retorting temperature between 800° and 1000°F. The hydrocarbons, water vapor, and gases are drawn off and the char or spent shale is separated from the ceramic balls in a revolving drum with holes in it. The ceramic balls are reheated in a separate furnace by burning some of the product gas. The coal pyrolysis primarily produces char, equal to about half the weight of the coal feed.[7] The liquid fuel and gas yields are very small, each amounting to about 10 percent by weight of the total of the three products. On the other hand, the oil shale pyrolysis yields crude shale oil as the principal product. Of course, the largest quantity of material evolved is the spent shale itself.

Foul water is always a significant part of coal pyrolysis products, principally because the coal is not dried and the coal moisture distills off in the pyrolysis. The foul water can be of the same order as the amount of liquid or solid hydrocarbons produced, depending largely on the moisture content of the coal (see

Table 3.3). The total water evolved is, however, made up not just of the coal moisture but also the moisture representing the oxygen in the coal, which was converted to water upon thermal decomposition of the organic components. Typically, one half to two thirds of the oxygen in the coal ends up as water vapor. For oil shale, the fraction of oxygen in the organic matter converted to water is about the same.[6] In shale processing, however, water of combustion represents only part of the water evolved. Another part comes from the thermal decomposition of the mineral carbonates present in the shale and the associated release of any water of hydration. McKee and Kunchal[8] have reported that when shale is pyrolyzed by burning a mixture of recycled product gas and air within a bed of shale, about half of the water evolved represents combustion product and about half free moisture and mineral carbonate thermal decomposition product. The foul water produced under such conditions can be as much as 30 to 40 percent by weight of the product oil.

Table 6.1

Reactor operating characteristics and products for processes to produce liquid and solid fuels from coal.

Process and Name	Reactor Pressure (psig)	Reactor Temperature (°F)	Principal Reactor Product
Hydroliquefaction			
Synthoil	2,000–4,000	850	Fuel oil
H-Coal	2,250–3,000	850	Synthetic crude or fuel oil and high-Btu gas
Solvent Extraction			
Exxon Donor Solvent	1,500–2,500	700–900	Naptha, fuel oils
Solvent Refined Coal	1,000–2,000	800–900	Solvent refined coal (pitch)
Liquid Hydrocarbon Synthesis			
Methanol (low pressure)	800	500	Methanol
Sasol (Fischer-Tropsch)	330–360	430–625	Liquid hydrocarbons
Pyrolysis			
TOSCOAL	~ atm	800–1,000	Char, small amount oil and gas
Garrett, COED	~ atm	1,100*	Char, pyrolysis oil, gas

*Staged from 550° to 1,550°F in the COED process.

Table 6.1 summarizes the general operating characteristics and products for the four different process types to convert coal to a liquid or solid fuel. From the brief process descriptions given, it is evident that no single generalized process train delineating the influent and effluent water streams can easily be drawn to represent all of the liquefaction and solid fuel conversion processes, as could be done for coal gasification. However, in hydroliquefaction and solvent extraction where the coal is dried, the amount of water evolved from the decomposition of the organic matter is of the same magnitude, and in the case of Solvent Refined Coal it may even be more than the theoretical amount of water required for the conversion (see Table 2.1). This means that the net process water requirement will not be large, if it is assumed that the foul process condensate that is generated is reused. The quantity of hydrogen that must be added to the reactor is therefore very much dependent on the coal properties. The principal input water is the steam required to manufacture the hydrogen, where part of the steam may be generated from cleaned up process condensate. Similar remarks apply to the pyrolysis of coal or shale if hydrotreating of the product is taken to be part of the process. In this case, net water productions are possible because the coal is not dried and because shale evolves water. Again, this depends strongly on the coal or shale properties, as well as the degree of hydrogenation. The process water streams for Fischer-Tropsch synthesis are largely those related to the manufacture of the medium-Btu gas, although some condensate is released from the synthesis reaction. As a general statement, therefore, the principal water streams in the production of liquid or solid fuels are those associated with gas manufacture, either hydrogen or a carbon monoxide/hydrogen gas. For the temperatures and process conditions of liquefaction, dirty foul condensate will be evolved, except for the synthesis process where the water quality is governed by the gas production.

The cooling water quantities per unit of heating value in the product fuel relate directly to the conversion efficiency, which is higher for coal liquefaction than for coal gasification. This follows from the fact that the hydrogen needed for liquefaction must be obtained via gasification. However, since less gasification is required, correspondingly less of the input fuel energy is needed. Assuming the process heat and power requirements to remain about the same, the conversion efficiency will be higher. Because of the way Fischer-Tropsch synthesis is carried out, the process has a low conversion efficiency. It would be expected to have cooling requirements approximately the same as those for the synthesis gas production.

6.2 Hydroliquefaction

In this section we show by example the process and cooling water requirements for the hydroliquefaction of coal. Our description is limited to the Synthoil and H-Coal processes. Only the detailed water streams for Synthoil are presented, since the differences between the two processes, from a water management point of view, are not large.

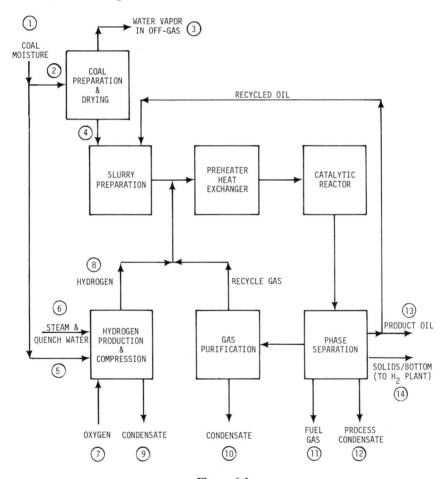

Figure 6-1
Water streams in a plant to produce oil from coal by hydroliquefaction with hydrogen production by gasification.

Figure 6-1 is a simplified block diagram of the important sections of a hydro-liquefaction plant with its principal water streams. Implicit in this design is the assumption that the hydrogen is manufactured by gasification of coal and solid residue to produce synthesis gas ($CO + H_2$), following which the CO is shifted with steam to produce hydrogen.

Synthoil

This process for the hydroliquefaction of coal to low sulfur fuel oil is being developed by the Pittsburgh Energy Research Center of the U.S. Department of Energy. An experimental plant that handles 0.5 ton/day of slurry is in operation at Bruceton, Pennsylvania;[9,10] a 10 ton/day process development unit will be completed in 1978.

The smaller pilot plant reactor is made of two interconnected stainless steel pipes, each 14.5 ft long and having a 1.1 inch inside diameter. The upper end of the first section is connected to the lower end of the second section by tubing so that it may be operated with one or both sections packed with catalyst and with the flow of fluid upward through each. Coal pulverized to 70 percent passing through 200 mesh screens (.003 inch opening) is dried to 0.5 percent moisture content and slurried to about 35 percent by weight in recycled oil. The slurry is mixed with hydrogen, preheated, and propelled by the gas in a two phase turbulent flow through the reactor, which is packed with pellets of cobalt-molybdenum catalyst. The reactor is operated at about 850°F at both 4,000 and 2,000 psig.

Upon leaving the reactor, gas is separated from the liquid. The hydrogen-rich gas is recycled to the reactor and makeup hydrogen is continuously added. Hydrogen sulfide and ammonia are purged. The oil is centrifuged from the suspended solids and some of it is recycled to slurry more coal.

An integrated plant design, including hydrogen production, is characterized by Fig. 6-1. The solids are a de-oiled char that is fed to the hydrogen plant for gasification along with the coal.[11] Several methods exist for the production of hydrogen including partial oxidation, steam reforming, and cryogenic separation. The method chosen here is partial oxidation in which synthesis gas is generated and then converted to hydrogen by the water-gas shift reaction (3.8). In steam reforming the hydrogen is produced from methane by the *steam reforming reaction,*

$$CH_4 + H_2O \rightarrow CO + 3H_2, \qquad (6.1)$$

followed by the conversion of the CO and steam to hydrogen by the shift reaction. The steam reforming reaction (6.1) is the reverse of the highly exothermic methanation reaction (3.9) and as such it absorbs a large amount of heat. The reaction is carried out at high temperature in a direct fired furnace. The type of hydrogen plant that is used with any given process will depend on a number of factors. These include the type of coal being processed (for example, whether the coal is high in moisture or sulfur), the product mix (for example, whether methane is produced that is not wanted), and the plant size. Partial oxidation is considered here for hydroliquefaction. Steam reforming will be considered in subsequent sections in connection with the Solvent Refined Coal process and the Sasol synthesis plant.

To illustrate the material balance and cooling requirements for the Synthoil process, consider the integrated design of Fig. 6-1 for a commercial plant to produce 50,000 barrels/day of low sulfur fuel oil with a specific gravity close to one and a heating value of about 17,600 Btu/lb. This is equivalent to a heating value output in the product fuel of about 3.1×10^{11} Btu/day and may be compared with about 2.4×10^{11} Btu/day from a 250×10^6 scf/day commercial size pipeline gas plant.

For the overall material balance, the required feed into the liquefaction reactor and the product flow out of the phase separation must be specified by experiment, as in the gasifier designs discussed in the last chapter. A summary of design criteria for the overall material balance is given in Table 6.2, where the stream numbers refer to Fig. 6-1. The data are based on experiments reported in Refs. 9 and 10 and design information in Ref. 11. To complete the information required, the coal must be specified. By way of example, we choose a relatively

Table 6.2
Design criteria for overall Synthoil material balance.

Material	Stream No.	Composition/Basis	Rate
Product oil	⑬	90 wt.% C, 8.5 wt.% H, 1.5 wt.% O, N, other	50,000 barrels/day
Carbon to reactor	②	5 barrels oil/ton C	833×10^3 lb C/hr
Char produced	⑭	6.2% C in coal	52×10^3 lb C/hr
Hydrogen requirement	⑧	4,700 scf/barrel oil	52×10^3 lb H_2/hr
CO from hydrogen plant	⑧	97 vol. % H_2, 3 vol.% CO	22×10^3 lb CO/hr
Coal oxygen	⑧	Convert CO to CO_2	13×10^3 lb O_2/hr
	⑪, ⑬	10% to gas and oil	—
	⑫	O_2 balance to condensate	—

Table 6.3
Coal used in the Synthoil example. New Mexico
subbituminous, Navajo/Farmington.

| | Weight Percent | |
	As-Received	Fed to Reactor
C	47.2	53.7
H	3.5	3.9
N	0.8	0.9
S	0.9	1.0
O	9.6	10.9
Ash	25.6	29.1
Moisture	12.4	0.5
Total	100	100
HHV (Btu/lb)	8,310	

low grade, high ash, low moisture content, New Mexico subbituminous coal. A typical composition and heating value for such a coal is shown in Table 6.3. Also shown in this table is the relative composition assuming that the slurried coal has been dried to 0.5 weight percent moisture. This moisture is assumed to remain in the product oil. Based on the carbon requirement shown in Table 6.2, the amount of as-received coal required for the reactor, but not including the amount needed for hydrogen production, is about 21,200 tons/day. With the preceding information an overall material balance, exclusive of the hydrogen production, can be calculated in a completely straightforward manner. Obtaining a complete hydrogen balance for the integrated plant design requires also that a material balance be carried out for the hydrogen production.

In Figure 6-2 a flow diagram is shown for the production of hydrogen by partial oxidation, along with the conditions needed to determine the water streams for a specified hydrogen output and given coal and char feed characteristics. Neither the design nor operating conditions are unique; however, they may be considered representative. The calculation of the material balance about the gas plant proceeds exactly as for an integrated gasification plant (without methanation), as detailed in Section 5.3. Enough coal is added to the char for the gasifier to be in thermal balance when the material balance is complete. This would be about 4,000 tons/day of as-received coal for the design considered. There were several necessary conditions used to make this calculation. In the gasifier off-gas, the mole ratio of H_2 to CO is 0.72. Since the hydrogen stream

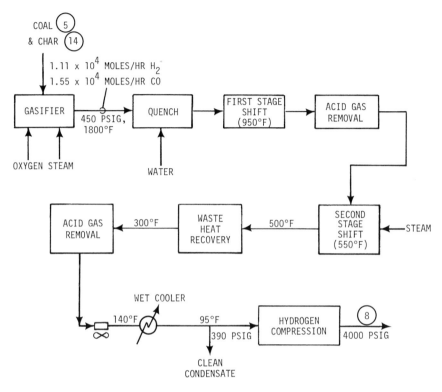

Figure 6-2
Flow diagram and operating conditions for hydrogen production by partial oxidation in Synthoil process.

⑧ contains 2.58 × 10⁴ moles H_2/hr and 0.08 × 10⁴ moles CO/hr, or a total of 2.66 × 10⁴ moles/hr, and since one mole CO yields one mole H_2 in a shift reaction, the total CO and H_2O in the off-gas must also be 2.66 × 10⁴ moles/hr and must be 1.11 × 10⁴ moles H_2/hr and 1.55 × 10⁴ moles CO/hr according to the specified mole ratios. Char, steam, oxygen, and coal are fed to the gasifier at rates determined by the simultaneous solutions of the carbon, hydrogen, and oxygen elemental balances and the thermal balance. For this high temperature gasifier, 80 percent of the steam feed is assumed to be decomposed. Hydrogen in the coal is first used up making the oxygen in the coal into water; only the surplus is available for reaction. Moisture in the coal passes unchanged into the gas.

The water equivalent hydrogen balance for the integrated plant is given in Table 6.4. The water streams in Table 6.4 are also expressed in gallons per

Table 6.4
Water equivalent hydrogen balance for Synthoil plant producing 50,000 barrels/day of 17,600 Btu/lb low sulfur fuel oil from 25,200 tons/day of as-received New Mexico subbituminous coal.

	10^3 lb/hr	gal/10^6 Btu*
In		
Moisture in as-received coal to liquefaction	219	2.04
Water equiv. of H_2 in as-received coal to liquefaction	549	5.10
Moisture in as-received coal to gasifier	42	0.39
Water equiv. of H_2 in as-received coal to gasifier	104	0.97
Total steam to H_2 production	265	2.46
Quench water to H_2 production	291	2.70
Total	1,470	13.66
Out		
From drying coal to liquefaction	211	1.96
Total dirty process condensate	318	2.96
Clean process condensate	70	0.65
Water equiv. of H_2 in product oil	572	5.32
Moisture in product oil	8	0.07
Water equiv. of H_2 and moisture in gas produced	284	2.64
Total	1,463	13.60

*In the product oil.

million Btu in the product oil. The heating value in the product oil is about 95 percent of the plant output, with the remaining 5 percent in fuel gas that is not used in the plant.

To determine the cooling water requirements, an estimate must be made of the unrecovered heat, its location, and its disposition.[12] The procedure closely parallels that outlined in Section 5.4 for a gasification plant. As we have emphasized throughout the book, for the plant to be in thermal balance, that part of the energy in the feed coal that does not leave as product oil or byproduct gas must leave the plant as unrecovered heat. For the conditions of the integrated conceptual design and for the coal used in our example, the thermal balance is given in Table 6.5. As expected, the conversion efficiency for hydroliquefaction is higher than for gasification. The efficiency of 80 percent may actually be somewhat high since only the principal energy loads in the plant were considered. However, this will not affect the cooling water requirement, because the additional losses are either

Table 6.5
Thermal balance for a 50,000 barrel/day
Synthoil plant.

Heating Value	10^9 Btu/hr
Coal feed	17.4
Product oil	(13.1)
Gas not used	(0.9)
Unrecovered heat	3.4
Overall conversion efficiency	80%

disposed of by air cooling (see Section 5.4) or are direct losses to the atmosphere, which have not been found.

In Tables 6.6 and 6.7, the unrecovered heat and its disposition are summarized. The heat loads to the dissolver heating and char de-oiling sections, which are unique to the process, were estimated from the design of Ref. 11, and the slurry pump requirements were based on pumping a 33 weight percent coal slurry. The heat loads for gasification are similar to those discussed in Section 5.4 for the Synthane process. They are based on corresponding compressor, turbine drive, gas removal, and water treatment requirements. The heat loads for electricity production and stack loss are also calculated in the same way as the Synthane example.

Table 6.6
Disposition of unrecovered heat lost directly and by dry cooling in a 50,000 barrel/day Synthoil plant.

	10^9 Btu/hr
Direct Loss	
Coal drying	0.235
Stack loss including char de-oiling	0.390
Electricity generated (27,000 kw), slurry pumps	0.115
Total Direct Loss	0.740
Dry Cooling	
Acid gas removal	0.465
Phase separation and other process streams	0.235
Total Dry Cooling	0.700

Table 6.7
Disposition of unrecovered heat lost by wet cooling and water evaporated
in a 50,000 barrel/day Synthoil plant.

Load	Heat Lost	Water Evaporated	
	10^9 Btu/hr	10^3 lb/hr	gal/10^6 Btu†
Turbine condensers*	0.885	632	5.87
Compressor interstage cooling	0.265	189	1.76
Other	0.810	579	5.38
Total Wet Cooling	1.960	1,400	13.01

*In arid regions or where water is expensive, 10 percent of cooling load on turbine condensers and 50 percent of load on interstage coolers is assumed wet cooled. Total evaporated water becomes 736×10^3 lb/hr or 6.8 gal/10^6 Btu.
†In the product fuel.

The liquefaction section of the plant does not account for the major heat load; rather it is the driving energy, as was the case with gasification. Specifically, cooling of the turbine condensers and compressor interstages accounts for almost 60 percent of the water evaporated. For this reason, the unrecovered heat disposed of by wet cooling is quite high (58 percent) compared to the gasification plants discussed in the last chapter. In arid regions, when a combined wet/dry cooling system is employed, the total evaporated water requirement can be reduced by about one half. Of the direct heat losses, the one most affected by the coal rank is the energy needed to dry the coal for liquefaction. This is a large load, directly proportional to the coal moisture, and an important factor in reducing the fuel conversion efficiency. The New Mexico coal used in the example had a relatively low moisture content.

The choice of wet or dry cooling for disposing of the various heat loads is based on the rules outlined in Chapter 4. The quantity of water evaporated is estimated on 1,400 Btu transferred per pound of water evaporated (see Table 4.1). However, estimates are not presented here for the quantity of water blown down. This does not affect the consumptive requirements, as long as this water is treated and reused or otherwise utilized.

The dirty condensate from the plant is about half from the liquefaction section and about half from the gasification section. (The quality of gasification water was discussed in Section 5.5.) Analyses of the dirty condensate from the liquefaction section have not been reported. Water analyses from a Solvent Refined Coal plant, presented in the next section, indicate a very dirty water. The Synthoil foul process condensate is expected to be comparable in contamination.

H-Coal

We conclude this section with a brief description of the H-Coal process.[13,14] This process, being developed by Hydrocarbon Research, Inc., employs a fluid bed catalytic reactor. Process development units in operation use reactors with an inside diameter of about 8.5 inches and 22 ft high, and handle about 2.5 tons/day of coal feed. A prototype demonstration plant is now being constructed with a hydrogenation reactor of 4.6 ft inside diameter. The plant will be capable of processing 250 to 750 tons/day of coal depending on whether synthetic crude or low sulfur fuel oil is the desired product (725 barrels/day of synthetic crude or 2,250 barrels/day of fuel oil will be produced). The commercial size reactors are anticipated to be about 10 ft or more in diameter.

The reactor shown schematically in Fig. 6-3 operates in the range of 2,250 to

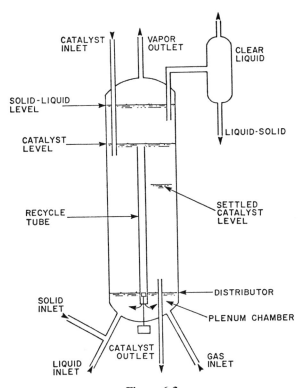

Figure 6-3
H-Coal fluidized bed reactor.

3,000 psig at 850°F and contains an active, bubbling bed of cobalt-molybdenum catalyst that can be added and withdrawn continuously to keep a constant level of activity. The active, bubbling, ebullient character of the bed has given rise to the name "ebullated bed." The bed is kept in a fluidized state by the upward passage of the slurry and gases. Higher pressure operating conditions are used when the desired product is a synthetic crude, and milder operating conditions are employed if a low sulfur fuel oil and a high-Btu gas are the desired products. The light hydrocarbon vapors are removed by an absorber, and ammonia and hydrogen sulfide are recovered from the remaining hydrogen gas, which is recycled to the reactor in the way described for the Synthoil process. The liquid stream is separated into light oil and heavy oil fractions by atmospheric and vacuum distillation, respectively, and a mixture of the oils is recycled to slurry the coal. The recycled oil mixture is dependent on the slurry oil composition desired. In commercial designs for the H-Coal process, hydrogen is assumed to be produced by the gasification of coal.[14] As noted at the beginning of this section, the procedure for estimating the water streams of an H-Coal plant with a corresponding product output would not be much different from that presented for Synthoil.

6.3 Solvent Extraction of Coal

Considered here are the process and cooling water requirements for the solvent extraction of coal, wherein the coal is partially dissolved in a solvent that exchanges hydrogen with the coal. Both the Solvent Refined Coal (SRC) and Exxon Donor Solvent (EDS) processes will be described, but only the results of detailed material and energy balances for the SRC process will be presented. The fundamental difference between the two processes is that the solvent in the EDS process is hydrogenated prior to mixing with the coal, whereas in the SRC process it is not. The SRC product is a solid at room temperature, while the EDS product is a liquid fuel. In broad outline, as represented by Fig. 6-1, both processes are quite similar to hydroliquefaction.

Solvent Refined Coal

We chose to present balances for the SRC process because it represents an example of the production of a clean solid fuel. The process is illustrated here with an integrated plant design in which the needed hydrogen is produced both by gasification and reforming. Figure 6-4 is a simplified block diagram of the integrated plant design and major water streams. In this design the undissolved

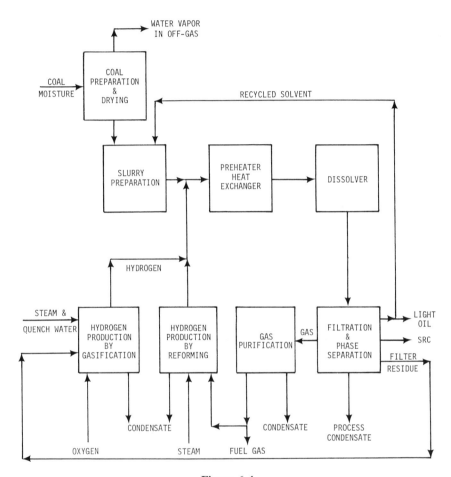

Figure 6-4
Water streams in a plant to produce solvent refined coal with hydrogen production by
gasification and reforming.

coal is used for the manufacture of hydrogen by gasification, and product gas is
used to manufacture some of the hydrogen by steam reforming.

The SRC process was developed by the Pittsburg & Midway Coal Mining Co.[6]
and two pilot plants are now in operation.[15,16] A 6 ton/day pilot plant is operated
by the Southern Services Co. at Wilsonville, Alabama, and a 50 ton/day pilot
plant is operated by Pittsburg & Midway at Fort Lewis, Washington. In these
plants raw coal is dried, pulverized, and mixed with a coal-derived solvent
boiling in the general range of 550° to 800°F. The coal-solvent slurry is pumped

together with hydrogen to from 1,000 to 2,000 psig and heated to from 800° to 900°F. Most of the carbonaceous material in the coal dissolves. The solution is filtered and the solvent is separated for reuse by vacuum distillation. The solvent refined coal so produced has a fusion temperature in the range of 300° to 400°F, generally has less than 0.2 weight percent ash and 1.0 weight percent sulfur, and has a heating value of about 16,000 Btu/lb.

The integrated plant design[21] used in this example was based on available pilot plant experiments (see Refs. 17 to 20). The integrated plant must necessarily include the manufacture of hydrogen, which in the pilot plant is bought. The integrated plant has been sized to yield 10,000 tons/day of solvent refined coal, or about 3.2×10^{11} Btu/day in the product fuel. This is very close to the output of a 50,000 barrel/day synthetic oil plant, which was shown in the previous section to be about 3.1×10^{11} Btu/day.

In the design illustrated here, all of the operating conditions from the coal preparation section to the phase separation and gas purification sections are based principally on the pilot plant designs. The pressure in the dissolver is taken to be 1,700 psia and the temperature to be about 850°F. The carbonaceous filter residue, a fine powder with a high ash content, is assumed to be gasified to produce as much hydrogen as possible. A Koppers-Totzek gasifier is used because of its

Table 6.8
Disposition of major coal elements in outputs from phase separation section of Solvent Refined Coal plant.

Coal Element	Fraction (%)	Output
C	70	SRC
C	14	Liquid hydrocarbons ($CH_{1.6}$)
C	5	Gaseous hydrocarbons ($CH_{3.7}$)
C	10	Filter residue
C	1	CO_2
N	100	SRC, Filter residue*, NH_3
O	100	Water, Filter residue†
S (organic)	65	H_2S
S (organic)	35	SRC, Filter residue
S (pyritic)	50	H_2S
S (pyritic)	50	Ash

*Coal N/C = Residue N/C; balance of N not in the SRC and residue is in NH_3.
†Coal O/C = Residue O/C; balance of O is in water.

performance on related materials. The general operating characteristics of this gasifier were discussed previously in Section 5.3. The remaining hydrogen needed in the process is manufactured by steam reforming of the product fuel gas. The reforming is carried out at 1500°F and 200 psig in a direct fired furnace followed by a shift reaction to convert CO. Equilibrium is assumed to be reached in the reformer and shift converter, and therefore the gas compositions can be calculated once the temperature is known. This, however, depends on the extent of reaction and involves the simultaneous solution of the heat and material balances around the units.

To carry out a material balance around the main sections of the plant outside of the hydrogen production sections, the disposition of the various elements in the feed coal must be known. Table 6.8 summarizes the outputs where the major coal elements appear from the filtration and phase separation section of the plant. The dispositions indicated are based on the pilot plant experience. Still to be specified are the feed coal and solvent refined coal composition.

In Table 6.9 two analyses are shown of solvent refined coal derived from eastern coals in the Southern Services Co. pilot plant.[21] Perhaps the most striking feature of these results is that there is actually a *decrease* in the hydrogen content in the solvent refined coal compared to that in the original coal, as shown by the molar representation. The main uses of hydrogen in the process are in the removal of oxygen from the coal by conversion to water, the removal of sulfur by

Table 6.9
Analyses in weight percent of solvent refined coals derived
from two eastern bituminous coals.

	Pittsburgh		Illinois	
	Coal	SRC	Coal	SRC
C	75.1	88.4	70.8	87.1
H	5.1	5.5	5.1	5.6
N	1.3	1.7	1.3	1.6
O	7.6	3.3	8.7	4.6
S	2.6	0.9	3.2	0.9
Ash	8.2	0.1	10.8	0.1
Molar Rep.	$CH_{0.82}$	$CH_{0.75}$	$CH_{0.86}$	$CH_{0.77}$
HHV(Btu/lb)	13,600*	16,000	12,900*	16,000

*From Dulong Formula (Eq. 3.1) including sulfur heating value.

conversion to hydrogen sulfide, and the hydrogenation to liquid and gaseous hydrocarbons of the carbon in the coal that is not converted to the solvent refined coal. From these analyses we could conclude that in principle the SRC process could be a net producer of water. Actually, whether the process is a net producer or consumer will depend on such things as the exact coal composition, how much water is left in the coal after drying, and details of the process design.

To compare the SRC process results with those from other processes for which western coals were used, Table 6.10 shows two typical western coals that have been studied in the SRC conceptual design.[21] The assumed solvent refined coal analysis is also given. From a number of reported analyses, the solvent refined coal does not appear to significantly differ from coal to coal.

Table 6.11 summarizes the total process water streams for the production of 10,000 tons/day of solvent refined coal. These streams were obtained from the appropriate material balances on the plant using the rules of Table 6.8 and the two coals of Table 6.10. There is a small net production of water with the Wyoming coal and a small net consumption with the New Mexico coal. This may be attributed to the higher oxygen fraction in the Wyoming coal. In general, the rules of Table 6.8 result in 60 to 70 percent of the oxygen in the coal being converted to water, although this may be low. In eastern coals, which have a lower oxygen content, up to 85 percent of the oxygen may be converted to water. The values of the net water produced shown in Table 6.11 may also be somewhat low since the coals were assumed to be dry. However, these will probably

Table 6.10
Assumed analyses in weight percent of solvent refined coal derived from two western subbituminous coals.

	New Mexico	Wyoming	SRC (assumed)
C	63.9	67.7	87.8
H	4.7	5.0	5.3
N	1.1	1.0	1.2
O	13.2	18.1	5.0
S	0.9	0.8	0.5
Ash	16.2	7.4	0.2
Moisture As-Received	16.3	19.9	0
Molar Rep.	$CH_{0.88}$	$CH_{0.89}$	$CH_{0.72}$
HHV(Btu/lb)	11,180	11,580	16,000

Table 6.11
Total process water streams for Solvent Refined Coal plant producing 10,000 tons/day
of solvent refined coal at 16,000 Btu/lb from western subbituminous coals.*

	Wyoming		New Mexico	
	10^3 lb/hr	gal/10^6 Btu†	10^3 lb/hr	gal/10^6 Btu†
In				
Steam and boiler feed water to process	451	4.06	389	3.50
Live steam to acid gas removal	264	2.37	250	2.25
Total	715	6.43	639	5.75
Out				
Foul water from phase separation section	212	1.91	146	1.31
Condensate from acid gas removal	264	2.37	250	2.25
Medium quality condensate from gasification section	163	1.47	160	1.44
Clean condensate from reforming section	86	0.78	48	0.44
Total	725	6.53	604	5.44
Net water produced	10	0.10	(35)	(0.31)

*Table 6.10. As-received coal; 23,116 tons/day Wyoming, 23,441 tons/day New Mexico.
†In the solvent refined coal.

contain about 0.5 percent moisture when fed to the dissolver and this water would then probably be recovered as dirty condensate.

Some information is available on the contaminants in the foul process water, and this is summarized in Table 6.12. The important point to be made is that the rate of production of ammonia, most of which may be assumed to dissolve in the condensed water and to leave with it, is very sensitive to the nitrogen content of the solvent refined coal. Data from the Fort Lewis pilot plant[21] show measured nitrogen contents of 13,000 mg/l with a standard deviation of 7,000 mg/l. Most of the hydrogen sulfide will also probably dissolve and come out in the foul water stream. Measurements at the Fort Lewis plant give a molar ratio for NH_3/H_2S of 2.0 with a standard deviation of 0.17. This corresponds to a sulfur content of 14,900 mg/l for a nitrogen content of 13,000 mg/l.

Table 6.13 is an overall thermal balance for the SRC plant based on an analysis of the energy requirements, the unrecovered heat, and its disposition. Only the

Table 6.12
Approximate analyses of foul process water
from Solvent Refined Coal plant.

	mg/l
Chemical Oxygen Demand	~30,000
Total Organic Carbon	~ 8,000
Phenol as C_6H_5OH	~10,000
Ammonia	15,800 ± 8,500
Sulfur	*

*Molar ratio NH_3/H_2S = 2 ± 0.17.

balance for the New Mexico coal is given since it differs little from that for the
Wyoming coal. These results are, however, more directly comparable with the
Synthoil example where a similar coal was used. Moreover, the SRC process is
particularly useful on this type of coal because of its high ash content. The
overall thermal balance shows the fractional energy requirement for the production of hydrogen by reforming, separate from the energy needed to drive the plant
and produce hydrogen by gasification. The overall plant conversion efficiency of
81 percent is just a bit higher than that for the Synthoil example, though the
difference is not significant. The efficiency may be considered representative for
this technology with a range between about 75 percent and the value of 81
percent shown.

Table 6.13
Thermal balance for Solvent Refined Coal plant producing
10,000 tons/day of solvent refined coal from New Mexico
subbituminous coal.

Heating Value	10^9 Btu/hr
Coal feed	18.28
Product SRC	(13.34)
Byproduct oil	(2.86)
Fuel gas and hydrogen produced	(2.08)
Fuel gas reformed	0.57
Fuel burned in plant	2.89
Unrecovered heat	3.46
Overall conversion efficiency	81%

Table 6.14

Disposition of unrecovered heat lost directly and by dry cooling in a Solvent Refined Coal plant producing 10,000 tons/day of solvent refined coal.

	10^9 Btu/hr
Direct Loss	
Coal drying, stack loss	0.67
Losses around filter and dissolver	0.36
Sensible heat of solvent refined coal	0.12
Electricity generated, slurry pumps	0.08
Other	0.06
Total Direct Loss	1.29
Dry Cooling	
Acid gas removal	0.25
Process dry cooling	0.81
Total Dry Cooling	1.06

The cooling water requirements, the unrecovered heat, and its disposition are summarized in Tables 6.14 and 6.15. Of the unrecovered heat, less than one third goes to evaporating water, as was the case in the gasification plant designs. The total requirement for evaporated water could be cut by more than half if this were deemed necessary by local conditions. Like the other synthetic fuel plants examined, the driving energy is the principal plant load. With an SRC plant, the important point is that the net water requirement for the plant is essentially only

Table 6.15

Disposition of unrecovered heat lost by wet cooling and water evaporated in a Solvent Refined Coal plant producing 10,000 tons/day of solvent refined coal.

Load	Heat Lost 10^9 Btu/hr	Water Evaporated 10^3 lb/hr	gal/10^6 Btu†
Turbine condensers*	0.64	457	4.11
Compressor interstage cooling*	0.14	100	0.90
Other	0.33	236	2.13
Total Wet Cooling	1.11	793	7.14

*In arid regions or where water is expensive, 10 percent of cooling load on turbine condensers and 50 percent of load on interstage coolers is assumed wet cooled. Total evaporated water becomes 331×10^3 lb/hr or 3.0 gal/10^6 Btu.

†In the solvent refined coal.

for cooling. Because of the relatively high plant efficiency, even this need is quite low, amounting to less than 2.3×10^6 gal/day in our example. If the plant has the equivalent of a 50,000 barrel/day fuel output, then this would in turn be equivalent to only about a barrel of water per "barrel" of fuel. This figure is small even if the solvent refined coal were considered to be a feedstock for refinery upgrading to alternative fuels. Of course, the water requirement estimate presupposes that any dirty water evolved in the process is treated for reuse within the plant and is not evaporated wastefully.

Exxon Donor Solvent

A solvent extraction technique for the production of liquid fuels, which has promise for commercial application, is the Exxon Donor Solvent (EDS) process.[2] The key element of the process is that the spent solvent is first catalytically hydrogenated in a separate fixed bed reactor prior to its use as a hydrogen donor solvent in the dissolver. With this intermediate step introduced, Fig. 6-1 could, for example, serve to represent an integrated plant design for this process. A 1 ton/day process development unit is currently in operation and a 250 ton/day pilot plant is projected with support from the U.S. Department of Energy.

The liquefaction reactors tested have included stirred tank, tubular flow with and without recirculation, and an ebullated bed (as described for H-Coal in Section 6.2), though the preferred reactor type in use and projected has not yet been specified. The reactor test conditions being studied include the following ranges: temperature, 700° to 900°F; pressure, 1,500 to 2,500 psig; solvent-to-coal ratio, 1.1 to 2.6, and residence time, 15 to 140 min. Recycle gas and light fuel products from the reactor are separated out in high and low pressure stages, respectively. The slurry product goes to vacuum distillation. The bottoms contain all the solid residue and high boiling hydrocarbons, of which little boils below 1,000°F. Part of the overhead product oil is the feed to the solvent hydrotreating system. The process is directed toward the production of naptha blending components and low sulfur fuel oil, with the relative amounts dependent on the liquefaction conditions. Integrated plant designs[2] indicate an overall conversion efficiency in the range of 65 to 75 percent, which would be at the lower end of the range suggested for hydroliquefaction processes. The process water requirements and quality may be expected to be comparable with those for hydroliquefaction.

6.4 Shale Pyrolysis Technologies

In discussing pyrolysis for the production of liquid and solid fuels, we first consider the pyrolysis of oil shale. This is because a liquid fuel is generally the main product of shale pyrolysis. In coal pyrolysis, large quantities of char are produced, thereby limiting the product's utility if the process is not combined with, say, gasification.

The pyrolysis or destructive distillation of shale to produce a crude shale oil, water vapor, and non-condensable gases is termed "retorting," and the reactor vessels in which this is carried out are called retorts. Two retorting options have been investigated extensively: mining followed by surface retorting and *in situ* retorting,[22,23] in which the shale oil is released by heating underground and then pumped to the surface (compare to *in situ* gasification described in Section 5.1). *In situ* retorting offers the possibility of eliminating the many problems associated with spent shale disposal, including the water required for this purpose, and of mining economically lower grade shales down to 10 gal/ton (see discussion in Section 3.2). At the present time, *in situ* processes are under development but they still present an array of problems, including environmental ones, and thus cannot yet be considered suitable for commercial operation. Nevertheless, the main qualitative features of *in situ* retorting, its relation to surface processing, and the principal water-related problems will be briefly described at the end of this section. Integrated plant designs and the related water quantity and quality characteristics will not, however, be outlined.

All surface processing operations involve the steps of mining, crushing, and then retorting to produce the shale oil. The product of the retorting is generally too viscous to be piped and too high in nitrogen and sulfur to be used as a synthetic crude for refining. The raw shale oil must therefore be put through an upgrading process, normally by hydrotreating. In addition, the spent shale from the retorting, equal to 80 to 85 percent by weight of the originally mined shale, must be disposed. In this section the emphasis will be on the retorting, which is at the heart of the surface processing operation. A flow diagram for the surface processing described is shown in Fig. 6-5.

Oil shale retorting is carried out at a temperature of around 900°F and for high grade shale requires from 200 to 300 Btu/lb,[6] depending on the retort temperature and the fraction of organic material in the shale. The heat for the retorting is provided by combustion of the product gas or residual carbon in the spent shale. Oil shale retorts are classified into two basic types, those that are directly heated

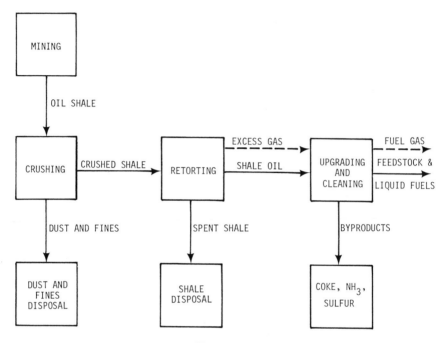

Figure 6-5
Flow diagram for surface processing of oil shale.

and those that are indirectly heated. In the directly heated processes, the heat is supplied by burning the fuel with air or oxygen within the bed of shale. In the indirectly heated processes, a separate furnace is used to heat a solid or gaseous heat carrier that is injected into the retort to provide the heat. By generating the heat indirectly, the pyrolysis gases are not diluted with combustion products and nitrogen from the air. We have already observed this distinction between directly and indirectly heated processes in coal gasification.

From the point of view of water management, the type of retort is quite important. One reason is that large quantities of water are evolved when shale is retorted, the actual amount depending on how the retort operates. A second reason is that some retorting processes require little or no cooling, while others must be cooled. The basis for the first reason lies in the fact that if the retort is heated directly, the water recovered from the retorting will be composed of the shale surface moisture, the thermal decomposition products of the carbonate minerals, and the water produced in the combustion. On the other hand, with indirectly heated retorts, the water in the combustion products is generally lost up the furnace stack and is not recovered.

In Section 6.1, results of McKee and Kunchal[8] were cited in which about half of the recovered water in a directly heated retort was said to be associated with combustion products. The data were from the Paraho process, discussed below, in which the surface moisture and thermal decomposition water were determined from laboratory tests. The water recovered over and above this amount in the direct fired retort was taken to be the combustion water. When the retort was heated indirectly, the water balance confirmed this division. A main reason for the difference in the cooling requirements in the directly heated retorts is that by countercurrent gas-to-shale flow, the sensible heat of the retorted shale is transferred to the entering gas, while the sensible heat of the product gas or oil is transferred to the feed shale. A result of the heat recovery is that the cooling loads are greatly reduced in these retorts, and the thermal efficiency is quite high. This is, however, not the only consideration, since the vapors and gases from directly heated retorts are diluted with combustion gases and hence require a greater absolute degree of cooling to condense the vapors and cool the gases. Generally, however, direct heating does reduce the load.

Table 6.16
Oil shale retorting technologies.

Name	Heating	Heat Transfer	Max. Feed Rate Operated (tons/day)
Union Oil	Direct	Gas-to-Solid	1,200*
Gas Combustion	Direct	Gas-to-Solid	~350†
Paraho	Direct	Gas-to-Solid	~550
Union Oil	Indirect	Gas-to-Solid	1,200*
Paraho	Indirect	Gas-to-Solid	~550
Petrosix	Indirect	Gas-to-Solid	2,500
IGT Hydroretorting	Indirect	Gas-to-Solid	25
TOSCO II	Indirect	Solid-to-Solid	1,000††
Lurgi-Ruhrgas	Indirect	Solid-to-Solid	**

*Last operated in 1958.
†Last operated in 1967.
**Commercial size plants for hydrocarbon cracking and coal devolatilization comparable to 5,000 tons/day.
††Last operated in 1972.

Oil shale retorting is a very old practice[6] and many processes have been put forward. At present, however, there are not a large number that can be considered well-developed and available for commercial production. In large measure this is a result of the volatile economics of production, which have alternately turned development on and off in the United States. In Table 6.16 those processes are listed that are well-developed or, as with the IGT Hydroretorting process, have considerable potential. Given in the table is the maximum feed rate at which the retort has been operated as of the summer of 1977. Not included here are multimineral processes such as the one being developed by the Superior Oil Co.,[24] where in addition to shale oil, minerals such as alumina and soda ash are recovered.

Direct (Gas-to-Solid) Heat Exchange

A direct retorting kiln that has served as the basis for other designs is the Gas Combustion Retort.[6] The development of this retort culminated in a nominal 150 ton/day pilot plant unit designed and operated by the Bureau of Mines at Anvil Points near Rifle, Colorado, from 1944 to 1955, at which time work was suspended.[25] An industry group modernized the 150 ton/day unit and operated it from 1964 to 1967 at shale feed rates as high as 500 lb/hr/ft^2 of kiln cross-sectional area, or about 350 tons/day.[26] Difficulties with clinkering (agglomeration) were a continual problem with the retort.

As operated, the retort was a rectangular, vertical, refractory lined kiln, 6 ft by 10 ft in cross section. The height from the kiln bottom to the shale feed tubes had a maximum value of about 35 ft. The feed tubes did, however, telescope and could be extended about a third of the way down the retort. Relatively coarse crushed shale was fed in at the top of the retort and flowed downward continuously by gravity as a moving packed bed. An optimum shale size was never determined, but in one of the more successful series of demonstration runs (Series C-1028)[26] the size of shale used was 1 in. × 2½ in. Direct gas-to-solid exchange of heat was supplied by internal combustion in the moving bed of shale.

The retort may be considered to be divided into four fictitious zones (that is, there are no physical or distinct separations) through which the shale passes. These zones are shown in Fig. 6-6. The shale first passes through the product cooling zone, where by countercurrent contacting it is heated almost to retorting temperatures by the gases rising from the retorting zone at the same time that the gases are cooled. The shale continues to move downward, entering the retorting

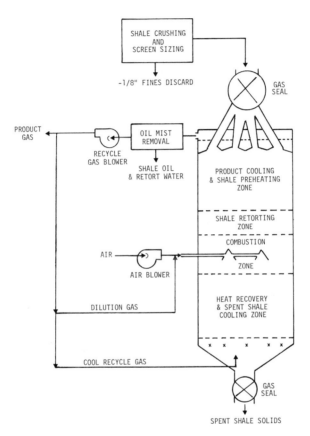

Figure 6-6
Gas combustion retorting zones.

zone where heat decomposes the kerogen into oil, water vapors, and noncondensable gases. The carbonaceous residue (char) from this reaction, which remains as part of the retorted shale, enters the combustion zone, where it is burned with air and the recycled low-Btu product gas to provide the heat for the process. The shale is heated to a maximum temperature of about 1400°F. The spent shale then moves down through a zone in the lower portion of the retort, where heat is removed from it by a rising countercurrent stream of recycled product gas. Besides giving the process a high thermal efficiency, this also enables the spent shale to be discharged at a relatively low temperature. For the successful demonstration series mentioned above the temperature averaged about 390°F.

The rising gas stream, which contacts the entering shale and preheats it, is a mixture of the liberated gases and vapors and the recycled product gas. Here, the basic process feature is that in the upper part of the retort the gas stream is cooled to a temperature below the dew point of the oil, which condenses as a fine mist or fog and is carried from the top of the retort with the gas stream. The temperature of the retort off-gas averaged 138°F for the same successful demonstration series noted. No wet cooling of the off-gas stream was required.

The gases leaving the retort pass through separators consisting of centrifugal blowers and electrostatic precipitators in series. These units separate out the shale oil together with the condensed retort water from the wet low-Btu product gas. The extent to which the retorted water is present in the off-gas as a vapor and as a mist depends on the system vapor pressure and temperature. The exact amount of water produced in the retort depends both on the operating conditions and on the characteristics of the oil shale.

In Fig. 6-7 the Paraho retort[27] is shown as it is operated in a directly heated mode. Like the Gas Combustion retort it employs countercurrent flow of gas and solids, with the shale moving continuously downward by gravity. The principal difference in the Paraho retort appears to be the feed distribution and shale discharge mechanisms, which are reported to maintain an even and continuous solids flow without clinkering. Although there are several levels at which product gas is injected, the primary fuel is the carbonaceous residue from the retorted shale.

Two retorts, one of 4.5 ft outside diameter and 60 ft high and one of 10.5 ft outside diameter and 75 ft high, were installed in 1974 at Anvil Points, Colorado, and are now operating. The retorts run in a direct mode, as well as in an indirect mode described below. The smaller retort, termed the pilot retort, has processed shale at rates above 700 lb/hr/ft^2 of retort cross-section (about 550 tons/day). A commercial scale retort is projected to be 104 ft high, have a 42 ft outside diameter, and process 11,500 tons/day of shale.

Jones reported[27] that in the direct mode the off-gas is removed at 140°F and the retorted shale is removed at 310°F. These temperatures are directly comparable with those cited for the successful Gas Combustion series. The yield of low-Btu off-gas from 28 gal/ton shale was given as 6,200 scf/ton of shale at 102 Btu/scf. Again these results are comparable with typical yields for the Gas Combustion retort.[6] The raw shale feed has a nominal size of $\frac{3}{8}$ in. $\times$ 3 in. and the carbon in the retorted shale is reduced to about 1.5 weight percent.

A retort whose principles are similar to those described, but whose operation is in the opposite direction, is the Union Oil retort.[6] It too has been run in both

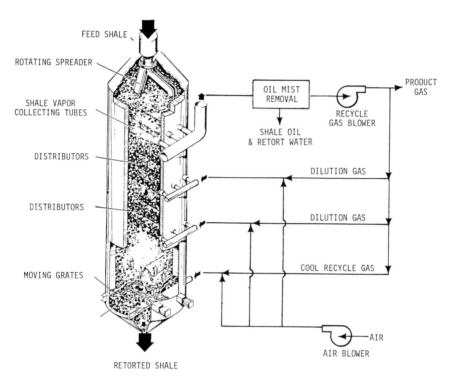

FEED SHALE

ROTATING SPREADER

SHALE VAPOR
COLLECTING TUBES

DISTRIBUTORS

DISTRIBUTORS

MOVING GRATES

RETORTED SHALE

OIL MIST
REMOVAL

RECYCLE
GAS BLOWER

PRODUCT
GAS

SHALE OIL
& RETORT WATER

DILUTION GAS

DILUTION GAS

COOL RECYCLE GAS

AIR

AIR BLOWER

Figure 6-7
Paraho retorting process—direct mode.

direct and indirect modes. In the directly heated retort, shale is charged into the lower and smaller end of a truncated cone and is pushed upward by a piston, referred to as a "rock pump," countercurrently against air caused to flow downward by suction blowers (see Fig. 6-8). The carbonaceous residue in the spent shale burns in the combustion zone where maximum temperatures of around 2,200°F are obtained. The oil is condensed on the cool, incoming shale and flows over it to an outlet at the bottom of the retort. This has the advantage over the Gas Combustion and Paraho retorts of not dripping oil products back into the hotter region, where they can be further cracked. Cooling water is also not required for this retort. In the late 1950's, the system was operated at up to 1,200 tons/day.

Indirect (Gas-to-Solid) Heat Exchange

The Paraho process is also operated in an indirect heat mode as shown in Fig.

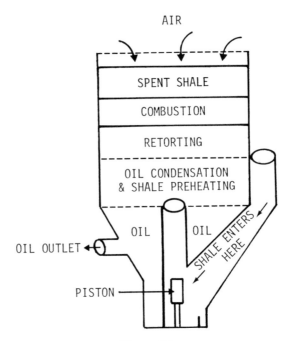

Figure 6-8
Retorting zones in Union Oil retort—direct mode.

6-9. In this mode the carbonaceous residue left in the retorted shale is burned in external heaters. The process heat to retort the shale is transferred to the raw shale at two levels by recycled product gas that has been circulated through the heaters. As a consequence of the external heat generation, the product gas remains undiluted by nitrogen from the combustion air, and thus a high-Btu gas is produced. Moreover, water is not recovered as a product of combustion but instead is lost up the heater stack. In this mode the off-gas is reported to leave at a temperature of about 280°F and the retorted shale to leave at a temperature of about 350°F.[27] A high-Btu gas yield of 500 scf/ton of shale is obtained with a heating value of about 885 Btu/scf from 28 gal/ton shale. The residual carbon in the retorted shale is reported to be about 3 percent by weight,[8] although the calorific value of the spent shale shown in the energy balance would indicate a value closer to 5 percent. The higher carbon content in the spent shale from the indirect process compared to the direct process, is due to the absence of air in the indirect retorting.

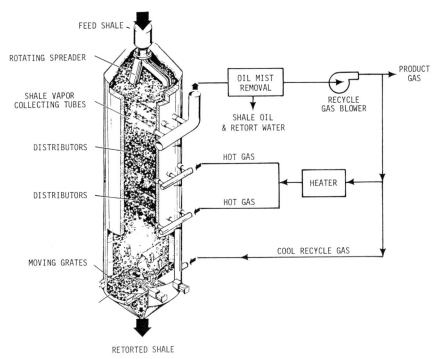

Figure 6-9
Paraho retorting process—indirect mode.

The Petrosix process developed by Brazil's national oil company employs a retort very similar in design to the Gas Combustion retort. It differs in operation in its use of indirect heating. Heated recycle gas rather than combustion air is injected into the shale bed at about the center of the retort, in contrast with the design shown in Fig. 6-6. A demonstration unit 18 ft in diameter and with a capacity to handle about 2,500 tons/day of shale has been operating since 1970, and a commercial facility is planned.

The Union Oil retort has also been operated in an indirect heat mode by using the low-Btu product gas from a direct heated retort to heat the high-Btu recycle gas in several indirectly heated retorts. The heated recycle gas was fed to the top of both the unfired retorts and the fired retort. In the unfired retort the heated recycle gas produced a maximum temperature of about 950°F, improving the thermal balance for the system as a whole by apportioning the recycle gas between the retorts.

A quite different retorting process that employs external heat generation is called hydroretorting. This process has been under development at the Institute of Gas Technology since 1954,[28,29] and a 1 ton/hr process development unit is now being operated. Its main feature is that the retorting is carried out in the presence of a moderate pressure hydrogen atmosphere of up to 500 psig. According to Ref. 29 the process extracts more hydrocarbon from all types of shale than does thermal retorting, possibly making it attractive for the refining of low grade shales.

In the retort, the crushed shale is fed in at the top and moves as a bed downward by gravity, through a reactor divided internally into three zones. The shale is preheated and prehydrogenated in the upper zone, hydroretorted in the middle zone, and the spent shale is cooled in the bottom zone. Cool hydrogen is fed to the bottom zone, recovering the sensible heat in the spent shale, and then is used to preheat the feed shale. A separate hydrogen stream that has been externally heated retorts the preheated shale in the middle zone from which the shale vapors are collected. By varying the reaction temperature, the product can be varied from mostly liquid when the temperatures are under 1200°F to a larger output of gas when the temperatures are over 1200°F.

Indirect (Solid-to-Solid) Heat Exchange

An indirect mode employing a solid-to-solid heat exchange is the TOSCO II process.[30] Work was begun in the mid-1960's on a 1,000 ton/day "semi-works" plant and carried on steadily through 1972. Intermittent testing has continued on a 25 ton/day pilot plant since that time and the process is considered ready for commercial operation. In fact, detailed design of a 50,000 barrel/day plant with six retorts each processing about 11,000 tons/day of 35 gal/ton raw shale was completed, and construction was to have begun in 1974, but the project was suspended indefinitely because of cost escalations and uncertainties concerning government policies.

In the process shown schematically in Fig. 6-10, crushed shale of minus $\frac{1}{2}$ inch size is preheated by pneumatically conveying the shale upward through a vertical pipe concurrently with hot flue gases from the ball heater. The flue gas is cooled during this process and the cooled gas is passed through a venturi wet scrubber to remove shale dust before venting to the atmosphere at a temperature of about 125° to 130°F.

The ball heater is a vertical furnace whose purpose is to heat up ceramic balls of about $\frac{1}{2}$ inch in diameter. After the balls are heated they are fed along with

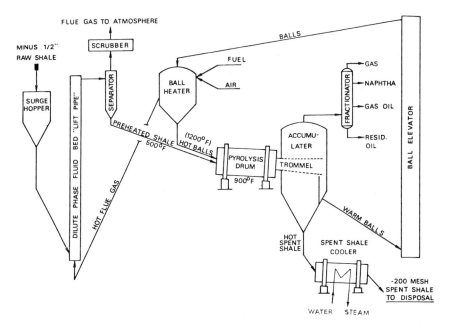

Figure 6-10
TOSCO II retorting process.

the preheated shale, which has been separated from the flue gas in settling chambers and cyclones, into a horizontal rotating kiln where the pressure is slightly above atmospheric. The mixture of balls and shale flows through the kiln, bringing the shale to a retorting temperature of about 900°F through conductive and radiative heat exchange with the balls. The resulting hydrocarbon and water vapors are drawn off and fractionated, leaving behind a mixture of balls and processed shale.

The ceramic balls are separated by size from the fine powdered spent shale by passage through a trommel (a heavy duty rotating cylinder with many small holes punched in its shell). Warm flue gas is used to remove residual dust from the ball circulation system. The dust is removed from the flue gas with a venturi wet scrubber. With a bucket elevator, the balls are then circulated back to the ball heater for reheating by burning some of the product gas.

The process shale is cooled in a rotating drum steam generator. One of the important features of the process is that the cooled, spent shale is then moisturized to approximately 14 percent moisture content in a rotating drum moisturizer, after which it is transported by a conveyor belt for disposal. The steam and processed shale dust produced in the moisturizing process are passed through another venturi wet scrubber to remove the dust before discharge to the atmosphere. The importance of the moisturizing is that addition of the water to the TOSCO-type processed shale, at a predetermined shale temperature, appears to lead to cementation of the shale after proper compaction. More importantly, the cemented shale appears to "freeze in" the moisture that has been added. We shall discuss this point further in the following chapter in connection with shale disposal problems.

A ton of 35 gal/ton oil shale is reported[30] to yield typically 1,652 pounds of processed shale, 250 pounds of oil, 70 pounds of high heating value gas, and 23 pounds of water. The gas from the retorting is composed only of the undiluted components from the oil shale itself and is therefore a relatively high-Btu gas. The yield of gas is about 915 scf/ton of shale and it has a heating value of about 775 Btu/scf. The residual carbon in the retorted shale is about 5 percent by weight.

Another solid-to-solid heating process is the Lurgi-Ruhrgas process, which has been developed commercially for the devolatilization of subbituminous coal and the cracking of liquid hydrocarbons.[31] Units suitable for the retorting of 5,000 tons/day of oil shale are presently in operation for cracking naptha. In the process, as proposed for adaptation to oil shale retorting, the heat carrier would be hot spent shale. The spent shale is mixed with the incoming raw shale in a sealed screw-conveyor, which acts as the retort. All of the material is discharged into a surge bin. The vapor stream is drawn off to a condenser to produce the gas and liquid fractions. Some spent shale is withdrawn from the surge bin as waste and the remaining solids are transferred to the end of a lift pipe. Here they are heated by combustion of the carbonaceous residue on the spent shale (and by burning supplemental fuel if needed). The hot solids are then used once again for retorting. Cooling, although not necessarily all-evaporative, will be required for the condenser. About 10 weight percent water is reported to be added to the spent shale for disposal.[31]

In Situ

The in situ procedure[22] involves fracturing the shale in place, igniting the fractured shale, feeding in air to sustain the combustion for pyrolysis, and pumping out the product shale oil and retort water that moves away from the combustion and retorting zones. In general, adequate flow paths are difficult to create for the gas and oil, even between closely spaced wells, because oil shale is not porous and does not lie in permeable formations. The result is that the gas flow is very difficult to control and the oil yields are low.

An alternative approach, known as modified *in situ,* is exemplified by the Occidental Oil process.[23] In this procedure a room is mined out by conventional techniques, removing about 15 to 20 percent of the shale deposit. The remaining shale is then exploded to expand it into the void volume. The resulting oil shale rubble constitutes the retort. After the retort is formed, connections are made to the surface for withdrawing the oil, retort water, and gas, and for pumping down air and recycle gas. Retorting is started at the top of the rubble pile by heating it with an outside energy source. After a certain amount of retorting, the energy source is turned off and the residual carbon in the shale is used as the fuel, with the combustion process sustained by injecting air. Recycle gas is also injected for control of the combustion. The combustion zone moves downward with the retorting zone below it and with the oil vapors condensing in the zone below the combustion region, analogous to the Union Oil process. The retorted oil and water flow by gravity to the bottom of the retort where they are collected in a sump and then pumped to the surface.

The processing criteria for the surface retort methods would apply in the same way to modified *in situ* processes. However, the main difference and advantage of such processes is that the solids disposal problem is substantially reduced. There are nonetheless a number of other environmental problems, among them treatment of the retort water, the migration of fluids during and after the retorting, surface thermal problems, and leaching of underground materials. The technology for *in situ* retorting is still relatively undeveloped and will not be discussed further.

6.5 Integrated Oil Shale Plants

Crude shale oils produced from the retorting processes described are classed as low to intermediate gravity, high nitrogen, intermediate sulfur, and waxy oils. Generally, these are more viscous and have a higher pour point (congealing temperature) than higher quality petroleum crudes. Although the oils from the various retorting processes differ somewhat, most of them must be cracked to lighter fractions to produce a pumpable oil. Moreover, nitrogen and sulfur levels must be reduced because of their deleterious effects on refinery catalysts and products. Most of the time the pour point is reduced by coking, and a pumpable oil with reduced nitrogen and sulfur levels is produced by hydrotreating.

To judge the water management problem associated with shale oil production on a basis comparable with coal liquefaction, comparable products must be considered. It is insufficient, therefore, to simply examine a retorting process by itself. Rather, it is necessary to consider a self-sufficient, integrated plant producing a high quality liquid fuel. For this purpose, we shall assume that the crude shale oil recovered from retorting is put through a pre-refining upgrading process to produce a low sulfur, low nitrogen, low pour, high quality, synthetic crude oil. Figure 6-11 diagrams the pre-refining upgrading process modified a bit from a suggested design for a commercial plant employing the TOSCO II retort.[30,32] The

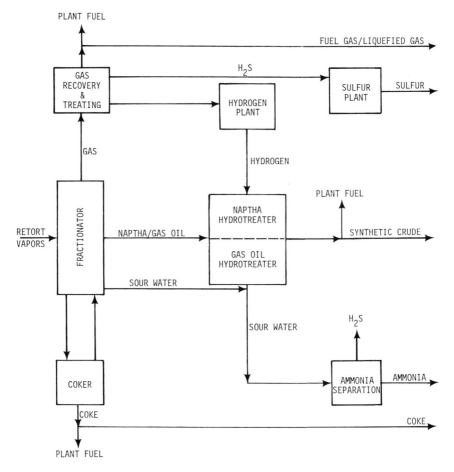

Figure 6-11
Shale oil upgrading plant.

vapor from the retort flows to a fractionation system where it is cooled to produce several liquid fractions and uncondensed gas. The heaviest oil fraction, comprising about a quarter of the oil, goes to a delayed coker. The gas, which for the indirect heated retorts is a high-Btu product and which for the direct heated retorts is a low-Btu product, is scrubbed to remove ammonia and sulfur. Part of the product gas or coke, or both, is routed to boilers for power generation. If the gas is a high-Btu one, then one of the byproducts will be a mixture of liquefied propane and propylene. The commercial plant produces its own hydrogen, which for the TOSCO II indirect retort is assumed to be by steam reforming. In addition, the plant is designed to include units for the separation of ammonia from hydrogen sulfide and the conversion of hydrogen sulfide to sulfur. We shall assume that an integrated shale oil plant has the essential features shown in Fig. 6-11.

In presenting material and energy balances for integrated plant designs, the Paraho Direct and Indirect processes and the TOSCO II process have been chosen to illustrate the three basic types of retorting technologies. Our choice is dictated not only by the commercial potential of the processes but also by the availability of published information. The unique feature of any oil shale plant is the retort. For this reason, the retort water streams, thermal balance, and process condensate quality are each presented separately. The upgrading step is essentially a refinery operation, and therefore the designs and balances will not be discussed in detail. Instead, process stream, cooling water quantities, and thermal balances derived from published data for proposed commercial scale plants will be presented.

To utilize the published design data and to convert from one design to another, the properties of the raw shale and the products must be specified. In Table 6.17

Table 6.17
Raw shale and product heating values and liquid product specific gravities.

Material	Property	Heating Value (Btu/lb)
Raw Shale	30 gal/ton	2,750
Raw Shale	35 gal/ton	3,208
Crude Shale Oil	0.928 spec. grav.	18,550
Synthetic Crude	0.825 spec. grav.	20,150
Liquefied Petroleum Gas	0.900 spec. grav.	21,200
Coke	—	13,850
Ammonia	—	8,620

the heating values and liquid product specific gravities are shown for the three designs discussed.[8,30]

There are different interpretations of the meaning of, say, a 50,000 barrel/day plant. The Paraho designs express the nominal output in terms of the crude shale oil production. The TOSCO design expresses the output as the sum of the upgraded liquid fuels. To have a uniform basis for comparing different retorting procedures and different grades of oil shale, we assume that the plant is designed to produce 50,000 barrels/day of synthetic crude plus any byproducts not utilized as plant fuel. From Table 6.17 the total heating value of the basic product fuel is 2.9×10^{11} Btu/day. When taken together with the byproducts, the output is directly comparable with the 3.1×10^{11} Btu/day from the 50,000 barrel/day fuel oil plants previously examined. Table 6.18 gives the net input and output quantities specified in the "100,000 barrel/day" Paraho designs[8] and in the "50,000 barrel/day" TOSCO design.[32] We have here converted the liquefied petroleum gas and coke outputs of the Paraho designs from heating value units to the units shown by using the values given in Table 6.17. The product differences seen in this table emphasize the need to clearly specify the meaning of the nominal plant size. In our use of the design data, all quantities will be scaled in the ratio of 50,000 barrels/day to the specified synthetic crude output of the design.

The water streams for retorting are necessarily determined from experiment. In Table 6.19 the results are summarized for the Paraho directly and indirectly heated processes[8] and for the TOSCO II process[30] (see also Ref. 12). The plant sizes from which the data have been scaled down (Paraho) or up (TOSCO II) to a 50,000 barrel/day synthetic crude output have been given in Table 6.18. For the

Table 6.18
Net input and output quantities as reported for integrated oil shale plant designs.

	Paraho Direct	Paraho Indirect	TOSCO II
Raw shale grade (gal/ton)	30	30	35
Sized shale (tons/day)	152,000	152,000	66,000
Purchased power (megawatts)	0	0	85
Synthetic crude (barrels/day)	87,000	76,000	45,000
Liquefied petroleum gas (barrels/day)	—	3,000	3,000
Coke (tons/day)	*	650	800
Ammonia (tons/day)	290	290	150
Sulfur (tons/day)	136	136	180

*Specified as sum of heat output of coke and low-Btu gas equal to 94×10^9 Btu/day.

Table 6.19
Retorting process water streams for oil shale plants producing 50,000 bbl/day
of synthetic crude in 10^3 lb/hr.*

	Paraho Direct	Paraho Indirect	TOSCO II
In			
Water addition to shale	28	32	50
Water into venturi scrubbers	—	—	172
Total	28	32	222
Out			
Water out in effluent sludge	—	—	53††
Water of retorting	272†	159**	83
Total	272	159	136
Net water produced	244	127	(139)

*In these plants 100×10^3 lb/hr = 1 gal/10^6 Btu of synthetic crude output.
†86% in retort product gas, 24% in crude shale oil.
**48% process condensate, 24% in retort product gas, 28% in crude shale oil.
††This water is assumed lost from the plant and so is not counted as a product.

product output considered here, the sized shale feed rate in the Paraho Direct and Indirect processes is 87,350 and 100,000 tons/day of 30 gal/ton shale, respectively. In the TOSCO II process the feed rate is 73,350 tons/day of 35 gal/ton shale.

One of the more interesting results is that the retort water output varies from a low of about 3.3 gal/ton of shale for the TOSCO II retort, to 5.2 gal/ton for the Paraho Indirect retort, to 9 gal/ton for the Paraho Direct retort. This represents between 14 and 45 percent by weight of the product synthetic crude. Expressed as a weight percent of the feed shale, it is between about 1.4 percent for the TOSCO II retort and 3.3 percent for the Paraho Direct heated retort. The value of 1.4 percent represents an amount of water very close to that which would be measured in a Fischer assay of the pyrolysis products.[6] This is not surprising since the TOSCO II retort is essentially a pure pyrolysis kiln. A weight percent of water in the Paraho Direct heated retort that is about double of that in the TOSCO II retort is also not surprising. The Paraho Direct retort is not a pure pyrolysis kiln, and the water of combustion is reported to account for about half of the retort water.[8] When the combustion water is subtracted out, the evolved water is then much more closely in line with what would be expected from pyrolysis alone. The quantity of retort water evolved in the Paraho Indirect process is intermediate between the other two processes, and this may be ascribed to the use of hot recycle gas as the medium for transferring heat.

Also shown in Table 6.19 is the net water production or consumption for the different retorting processes. It should be emphasized that the TOSCO II process shows up as a net consumer compared to the Paraho processes, because the design uses wet venturi scrubbers for off-gas cleaning. Other alternatives are possible. In any case, water requirements cannot be characterized by the retorting water alone but must be based on the needs for the integrated plant. The absolute process water consumption or production for both retorting and upgrading is but a small quantity compared to that for cooling or for spent shale disposal. However, the water that is produced is very dirty. This situation is similar to that prevailing for the coal gasification and liquefaction plants, although in the surface retorting oil shale plants a large quantity of water is also required for spent shale disposal.

Little information has been published on the quality of the retort water. Table 6.20 summarizes an analysis of Gas Combustion retort water taken from Ref. 6 (see also Ref. 22) and an analysis of TOSCO II retort water representing a composite from data in Refs. 30 and 33. The Gas Combustion retort water is probably quite similar to that from the Paraho directly heated retort. Both waters are seen to be very high in ammonia and sulfur (as hydrogen sulfide) and to be quite dirty. From the value for the total carbon content, the organic carbon content is most probably much higher in the Gas Combustion retort water[34] than in the TOSCO II water. This can also be explained as a consequence of the

Table 6.20
Analyses of retort water from Gas Combustion and TOSCO II
processes, mg/l (except pH).

Gas Combustion*		TOSCO II	
pH	8.8	pH	~8–9
Total Carbon	18,500	Carboxylic acids†	1,000–2,000
Carbonate	14,400	Phenols	50
Ammonia	12,400	Ammonia**	15,800
Chloride	5,400	Sulfur	4,000
Sulfate	3,100		
Sulfur (nonsulfate)	1,900		

*Separated from retort oil.
†82 wt.% straight chain, 18 wt.% isomeric and unidentified.
**Nitrogen as ammonia.

difference in retorting procedures, the TOSCO II products being more representative of those from pure pyrolysis. What is most surprising in the analysis of the TOSCO II water is the very low content of phenols, with the organic acids overwhelmingly carboxylic acids. Cook[33] notes that the straight chain carboxylic acids that predominate in the water are highly biodegradable, an important feature when considering water treatment.

In Table 6.21 a summary is given of the water streams for the upgrading sections of the plants considered. The makeup water represents water consumed in large part in the hydrogen plant, as well as water consumed in gas treating, in coking, and in other process steps. The foul water, from which ammonia and hydrogen sulfide are stripped out, is made up principally of the retort water and the foul water from the gas treating unit and the coker. (The TOSCO II values were not obtained directly from Ref. 30, which contains some misprints in the water flow diagram, but from a corrected water system flow diagram supplied by the Oil Shale Corp.) Only the lumped streams reported for the Paraho processes are shown in the table. The statement that the foul water is for reuse does not necessarily mean that it is for reuse within the retorting or upgrading units. Indeed, most designs have envisaged using this water for spent shale disposal, about which we shall speak in the next chapter.

Summarized in Table 6.21 are the net water consumptions for the upgrading

Table 6.21
Upgrading water streams and net water consumed or (produced) in retorting and upgrading sections for oil shale plants producing 50,000 barrels/day of synthetic crude in 10^3 lb/hr.*

	Paraho Direct	Paraho Indirect	TOSCO II
In			
Retort water	272	159	83
Makeup water	378	433	444
Total	650	592	527
Out			
Foul water for reuse	439	350	266
Boiler blowdown	83	95	119
Total	522	445	385
Net water consumed	128	147	142
Net water consumed in retorting and upgrading	(116)	20	281

*In these plants 100×10^3 lb/hr = 1 gal/10^6 Btu of synthetic crude output.

operations. It is rather interesting just how close the requirements are for the Paraho and TOSCO plants, despite the fact that the values were drawn from different plant designs. This supports the point made earlier in the section that the pre-refinery upgrading process would be similar in its essential features for the same product output and mix. Also shown in the table is the net water consumption or production for the combined retorting and upgrading steps. The Paraho direct heated process is seen to be a net producer of water, a consequence of the carry over of water evolved in the retorting section. However, we again emphasize that the absolute value of these net waters are small compared to both the cooling water and shale disposal water.

Shown in Table 6.22 are the thermal balances made for the specified outputs of the three processes. The retorting heat and retorting power requirement for the Paraho processes have been scaled from Ref. 8. The retorting heat for the TOSCO II process is estimated at 260 Btu/lb. We have included in the heating value of the crude shale oil sensible heat estimated at about 100 Btu/lb. Also shown in the table are the retort conversion efficiencies. As would be expected, the highest efficiency is attained by the direct combustion process where no intermediate medium is used to transfer heat for the pyrolysis. The higher efficiency of the TOSCO retort compared with the Paraho Indirect retort is a consequence of the fact that in the TOSCO kiln, the heat is transferred by direct solid-to-solid conduction and radiation. This mode of heat transfer is more efficient than the gas-to-solid heat transfer used in the Paraho retort.

Although the retorting efficiencies are relatively high, they mean little by themselves since the product of interest is not crude shale oil but an upgraded

Table 6.22
Retort thermal balances for 50,000 barrel/day oil shale plants.

Heating Value	10^9 Btu/hr		
	Paraho Direct	Paraho Indirect	TOSCO II
Sized shale feed	20.0	22.9	19.6
Retorting heat	—	1.8	1.6
Power for retorting*	0.4	0.5	0.5
Crude shale oil	(14.5)	(16.6)	(14.3)
Untreated product gas	(3.1)	(1.9)	(2.2)
Unrecovered heat	2.8	6.7	5.2
Overall conversion efficiency	86%	73%	76%

*10,000 Btu/kwh (34% conversion efficiency).

Table 6.23

Thermal balances, unrecovered heat removed by wet cooling, and water evaporated in 50,000 barrel/day oil shale plants.

Heating Value	10^9 Btu/hr		
	Paraho Direct	Paraho Indirect	TOSCO II
Sized shale feed	20.0	22.9	19.6
Purchased electricity*	—	—	0.9
Power to mine and size*	0.3	0.3	0.2
Synthetic crude	(12.1)	(12.1)	(12.1)
Liquefied gas	—	(0.6)	(0.9)
Coke	(2.3)†	(0.5)	(1.0)
Ammonia	(0.1)	(0.1)	(0.1)
Unrecovered heat	5.8	9.9	6.6
Overall conversion efficiency	71%	57%	68%
Fraction of unrecovered heat to evaporate water	28%	19%	18%
Water evaporated for cooling (10^3 lb/hr)	1,160	1,330	850

*10,000 Btu/kwh (34% conversion efficiency).
†Heating value of coke and low-Btu gas.

synthetic crude. Table 6.23 gives the thermal balances for the integrated plants. It can be seen that the Paraho Indirect process has a quite low conversion efficiency, while the efficiency of the Paraho Direct process is comparable with coal liquefaction. The efficiency for each plant as a function of its corresponding retort efficiency is plotted in Fig. 6-12. Also shown in the figure is a line labelled "refinery efficiency." This is arbitrarily set at a level of 75 percent, which may be considered approximately representative for the upgrading operation alone. The curve drawn between the points shows that at high retorting efficiencies the contribution of the retorting inefficiencies to the reduction of the overall plant efficiency is small, as would be expected. On the other hand, at low retorting efficiencies, comparable to that of the "refinery," the retorting inefficiency has a weight more equal to that of the refinery, and the plant efficiency decreases more or less linearly with the retort efficiency. In viewing these results, however, keep in mind that only three points have been used to plot the curve in Fig. 6-12, so that its usefulness is probably more qualitative than quantitative.

Also shown in Table 6.23 is the water evaporated for cooling in the plant designs. When the water evaporated is expressed as a fraction of the unrecovered

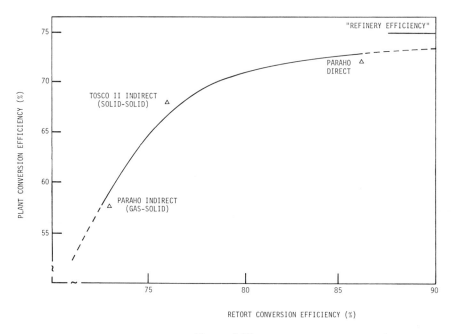

Figure 6-12
Oil shale plant conversion efficiency as a function of retort conversion efficiency.

heat removed, assuming 1,400 Btu transferred per pound evaporated, the result for the Paraho Direct process is within the range of what we have shown previously for the coal conversion plants studied. The fraction of heat used to evaporate water in the indirect processes is somewhat lower. This may be explained by the fact that in the indirect processes part of the unrecovered heat is lost up a furnace stack, which is not lost that way in the direct processes.

6.6 Coal Pyrolysis

Without additional processing, the major product of coal pyrolysis will be char with smaller amounts of oil and gas. Among the several coal pyrolysis processes described in Section 6.1, this was most evident in the TOSCOAL process, which is a direct application to coal pyrolysis of the oil shale technology outlined in the preceding two sections. Since synthetic fuels and not char production are our interests, coal pyrolysis will be illustrated by a process that has one part designed to produce both char and a synthetic crude oil, the latter obtained by hydrotreating the pyrolysis oil. In the companion part of the process, the char is gasified to produce pipeline gas, as outlined for coal in Section 5.1. The part of the process to produce char and a synthetic crude is called the Char-Oil-Energy Development process or COED process.[6,35] When combined with the companion process to produce synthetic gas, it is the COGAS process.[6]

The COED process to produce synthetic crude is basically similar to the production of synthetic crude from oil shale by pyrolysis and hydrotreating. The process was conceived with the intention of converting coal to more useful fuels using the simplest and most developed technology. It has been under development since 1962 by the FMC Corporation, which operated a 36 ton/day pilot plant in Princeton, New Jersey, that recently has been dismantled. The COED process by itself cannot stand alone, but rather represents a part of an integrated plant as may be represented by the COGAS process. Under sponsorship of the U.S. Department of Energy, a COGAS demonstration plant is being designed for construction in southern Illinois to produce 18×10^6 scf/day of pipeline gas and 2,400 barrels/day of low sulfur synthetic crude oil from a feed of 2,200 tons/day of coal.

A flow diagram of the COED process is shown in Fig. 6-13. Coal is crushed, dried, and then heated to successively higher temperatures in a series of fluidized bed reactors operated at low pressure (5 to 10 psig). In each fluidized bed, pyrolysis liberates a fraction of the volatile matter of the coal. The temperature of each bed is just below the temperature at which the coal would agglomerate and plug the bed. Once the coal is partially devolatized in one reactor, it can then be heated further in the next reactor. Typically, four stages are used, operating at 550°, 850°, 1000°, and 1550°F with a range of ± 50°F. The number of stages and the operating temperatures vary with the agglomerating properties of the coal. Heat for the process is generated by burning char in the fourth stage with a steam-oxygen mixture and then using hot gases and the hot char from the fourth stage to heat the other vessels. The pyrolysis is therefore in the category, defined for oil shale retorting, of an indirectly heated process with gas-to-solid heat exchange.

The volatile products released from the coal in the fluidized bed reactors pass to a product recovery system where oil, water, and gases are separated. The pyrolysis oil, tarry and laden with solids and sulfur, is filtered to remove the solids. Like crude shale oil, the solids-free heavy pyrolysis oil cannot be used directly as a refinery feed stock because of its high nitrogen and sulfur content. The oil is upgraded to a higher quality synthetic crude by catalytic hydrogenation in a fixed bed reactor where sulfur, nitrogen, and oxygen are removed.

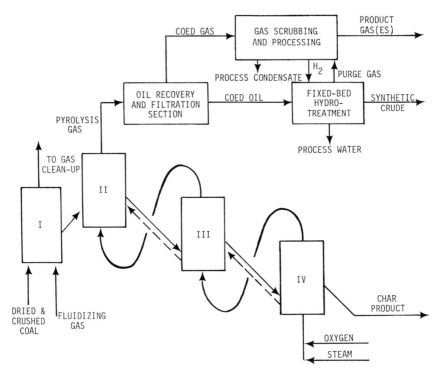

Figure 6-13
COED coal pyrolysis and hydrotreating.

In a commercial plant, the medium-Btu product gas not used to make hydrogen or not used for plant fuel would be upgraded to pipeline gas. Some of the char product may be burned to meet plant fuel requirements. However, with a high sulfur feed coal, the char usually contains too much sulfur to meet emission standards.

A detailed material balance has been made for a COED pyrolysis and upgrading plant.[36] The coal feed rates and net product outputs for this plant are given in Table 6.24. They have been scaled to an approximate 50,000 barrel/day output by simply doubling all of the rates in the reported design. The product outputs were taken from Ref. 37, where a detailed thermal balance for the design of Ref. 36 showed that for a self-sufficient plant all of the product gas and even some of the char (about 1,450 tons/day) would have to be used for plant fuel. For this reason, no product gas appears as an output in Table 6.24. However, some difference of opinion exists on this point, and as an example a net output of

Table 6.24
Net input and output rates and heating values for
COED pyrolysis and hydrotreating plant.

Material	Rate (tons/day)	Heating Value (Btu/lb)
Coal (dried)	48,000	12,420
Synthetic crude*	7,892	19,100
Char	23,568	11,700
Sulfur	780	3,990

*Specific gravity = 0.904.

product gas is indicated in Ref. 38. The heating value of the total plant output is about 8.6×10^{11} Btu/day but that of the synthetic crude alone is 3×10^{11} Btu/day. This may be compared with 3.1×10^{11} Btu/day for the heating value of the fuel oil output from the liquefaction plants examined, or with 2.9×10^{11} Btu/day for the synthetic crude output from the oil shale plants considered. As envisaged for commercial operation, the pyrolyzer vessels are each 60 to 70 ft in diameter. For the coal feed indicated, a total of sixteen such pyrolyzers in four trains would be required.

From the material balance of Ref. 36 an approximate water equivalent hydrogen balance has been made. The balance for the pyrolysis section is given in Table 6.25, and that for the upgrading section separately in Table 6.26. The net water consumed represents the difference between stream ① and the sum of streams ② and ③, which is quite small and comparable with that for indirect oil shale retorting. A most interesting result follows from Tables 6.25 and 6.26 by balancing the molecular hydrogen in streams ④ to ⑧. In particular,

$$④ + ⑦ \approx ⑥ + [⑧ - ⑤].$$

This implies that the plant has a sufficiency of hydrogen, provided none of the product gas is used to drive the plant. In this case a hydrogen purification section would be required but it would not be necessary to have a special plant to produce the hydrogen for the hydrotreating step, so that there would be no net consumption of water. Indeed, according to Table 6.26, water would be produced. In fact, this is not the case, and in the FMC commercial design[38] some hydrogen is manufactured by steam reforming a part of the cleaned pyrolysis gas. Still, the net water consumed in the upgrading section should be small and considerably less than that for shale oil hydrotreating listed in Table 6.21.

Tables 6.25 and 6.26 show three major sources of process water, represented

Table 6.25
Water equivalent hydrogen balance for pyrolysis section of COED plant
producing 50,000 barrels/day of synthetic crude and 24,000 tons/day
of char from 48,000 tons/day of dry Illinois bituminous coal.

Stage/Stream		10^3 lb/hr	gal/10^6 Btu*
	In		
I	Moisture in coal to pyrolysis	252	2.42
I	Water equiv. of H_2 in coal	1,416	13.60
IV/①	Steam to pyrolysis	1,348	12.95
	Total	3,016	28.97
	Out		
I	Moisture in product gas	72	0.69
I/②	Process condensate	372	3.57
I	Water equiv. of H_2 and hydrocarbons in product gas	24	0.23
II	Moisture in product gas	114	1.10
II/③	Process condensate	948	9.10
II/④	Water equiv. of H_2 in product gas	446	4.28
II	Water equiv. of hydrocarbons in product gas	474	4.55
II/⑤	Water equiv. of H_2 in COED oil	550	5.28
	Total	3,000	28.80
	Net water consumed in pyrolysis	28	0.27

*In the synthetic crude.

by streams ②, ③, and ⑨. In Table 6.27 representative analyses of these waters are shown.[38] The condensate from the first stage pyrolysis section is quite clean. On the other hand, the second stage foul process condensate is quite dirty. Its quality is similar to that from Gas Combustion retorting (Table 6.20) or from Synthane gasification (Table 5.24). The process water from the hydrotreating section is laden with hydrogen sulfide and ammonia, and the dissolved organic contamination is comparable with that of the second stage process condensate.[38]

Our presentation of the COED process is concluded by showing in Table 6.28 an overall thermal balance for the plant. No cooling water estimate is available since a COED plant would normally not be designed to stand alone. A rough estimate could be made of the cooling water requirement to produce only the synthetic crude, by assuming that 17.4×10^9 Btu/hr in the coal feed goes to

Table 6.26
Water equivalent hydrogen balance for upgrading section of COED plant producing 50,000 barrels/day of synthetic crude and 24,000 tons/day of char from 48,000 tons/day of dry Illinois bituminous coal.

Stream		10^3 lb/hr	gal/10^6 Btu*
	In		
	Water equiv. of H_2 in COED oil	550	5.28
⑥	Water equiv. of H_2	738	7.08
	Total	1,288	12.36
	Out		
⑦	Water equiv. of H_2 in purge gas	396	3.80
	Water equiv. of hydrocarbons in purge gas	50	0.48
⑨	Process water	66	0.63
⑧	Water equiv. of H_2 in synthetic crude	650	6.24
	Total	1,162	11.15

*In the synthetic crude.

producing crude at 72 percent efficiency. This represents 35 percent of the feed heating value, the same percentage that the synthetic crude heating value bears to the total product heating value. Also, of the 4.9×10^9 Btu/hr of unrecovered heat, 25 percent of it is assumed to be dissipated by wet cooling at 1,400 Btu transferred per pound of water evaporated. In that case, the water evaporated is 875×10^3 lb/hr or 8.4 gal/10^6 Btu of synthetic crude.

Table 6.27
Analysis of process waters from pyrolysis and hydrotreating sections of a COED plant using Illinois #6 bituminous coal, mg/l (except pH).

	Stage I	Stage II	Hydrotreating
pH	3.6	8.8	9.3
Suspended Solids	4,900	10,900	—
Entrained Oil	—	0–5,000	—
Carbon	—	—	8,000
Phenol	0	3,800	—
Ammonia*	600	11,300	60,700
Sulfur	700	1,800	87,000

*Nitrogen as ammonia.

Table 6.28
Thermal balance for a COED pyrolysis and
hydrotreating plant producing 50,000
barrels/day of synthetic crude and 24,000
tons/day of char from 48,000 tons/day
of dry Illinois bituminous coal.

Heating Value	10^9 Btu/hr
Coal feed	49.68
Synthetic crude	(12.56)
Char	(22.98)
Sulfur	(0.26)
Unrecovered heat	13.88
Overall conversion efficiency	72%

6.7 Liquid Hydrocarbon Synthesis

The last procedure to consider for the liquefaction of coal is liquid hydrocarbon synthesis, in which synthesis gas (CO + H_2) is reacted in the presence of a catalyst to form hydrocarbon vapors that are then condensed to liquid fuels. The procedure is termed Fischer-Tropsch synthesis when hydrocarbon compounds are the principal products, as distinct from the synthesis of oxygenated hydrocarbons such as methanol. Our discussion will focus on Fischer-Tropsch synthesis. The principles for both processes are, however, the same as illustrated in Fig. 3-1 and discussed in Section 3.5.

The main drawback of the Fischer-Tropsch process is its poor overall conversion efficiency, a result of the combination of two relatively inefficient processes—gasification and Fischer-Tropsch synthesis. A recent study by the R. M. Parsons Company[39] of a Fischer-Tropsch complex to produce oil and gas suggests, however, that with ''second generation'' gasifiers and synthesizers an overall efficiency of about 70 percent may be attainable. Apart from the economic penalties, our concern with low efficiency is related to the higher cooling water requirements. Despite the drawbacks of the process, it is the basis of the only fully operational size facility in the world for the production of motor fuels from coal. This facility is the Sasol plant of the South African Coal, Oil and Gas Corporation which has been in operation since 1955. At this plant both motor fuels and a high-Btu gas are produced from coal. Moreover, a second generation plant is now being built for operation in 1981 that will have an output of about

50,000 barrels/day of gasoline.[40] Its completion will represent a major achievement in the commercial production of synthetic fuels.

At the presently operating Sasol plant, about 7,900 tons/day of 10.7 weight percent moisture coal, with a high ash (35 weight percent) and low sulfur content, are first gasified to produce a medium-Btu gas. The heating value of the medium-Btu gas is about 400 Btu/scf and that of the coal 8,380 Btu/lb dry. The gas then undergoes a Fischer-Tropsch synthesis to yield about 93 tons/day of liquid fuels (about 8,000 barrels/day of motor fuels with an average specific gravity of 0.66) and 60×10^6 scf/day of 500 Btu/scf tail gas.[41]

Gasification is by the Lurgi system, described in Section 5.3. There are 13 oxygen-blown, 12 ft diameter gasifiers operating at 400 psig and producing some 330×10^6 scf/day of raw gas. The gas is then quenched and scrubbed for removal of particulate matter, tar, and ammonia, following which H_2S and CO_2 are removed by the Lurgi Rectisol system (see Section 3.6).[42] The heart of the conversion process is the Fischer-Tropsch synthesis. At the Sasol plant there are two kinds of synthesizers operating in parallel. The Arge system, which uses a pelletized iron catalyst in a fixed bed tubular reactor, and the Kellogg-Synthol system, which uses a promoted iron catalyst powder in a circulating fluidized bed. The fluid bed system has a higher output per reactor volume and is useful for the production of lighter hydrocarbons, while the fixed bed system is suitable for a significant contribution of higher molecular weight hydrocarbons.

The 50,000 barrel/day Sasol plant now under construction will be used to illustrate the water streams in Fischer-Tropsch technology. The design output of this plant is about 476×10^3 lb/hr of 20,700 Btu/lb fuel, principally gasoline. This is equivalent to 2.36×10^{11} Btu/day or about the same heating value output as that from the 250×10^6 scf/day pipeline gas plants considered in Chapter 5, but less than the 3.1×10^{11} Btu/day in the output of the hydroliquefaction plants. The plant incorporates many improvements gained from its predecessor. Figure 6-14 is a simplified flow diagram of the plant.[40] It differs from the existing version in that only the Synthol circulating fluid bed reactor has been incorporated, because the main purpose is to produce gasoline. The plant will contain ten Synthol reactors, each with a design capacity of about 2.5 times the existing ones, or about 5,000 barrels/day capacity compared with 2,000 barrels/day.

Figure 6-15 is a schematic of the Kellogg-Synthol reactor.[41] Feed gas enters at 320°F and 300 psig and mixes with and entrains the catalyst which is at 635°F. The suspension enters the fluidized bed reaction section where the Fischer-Tropsch and water gas shift reactions proceed. Heat liberated by the exothermic reactions is removed by oil coolers and used for steam raising and preheating

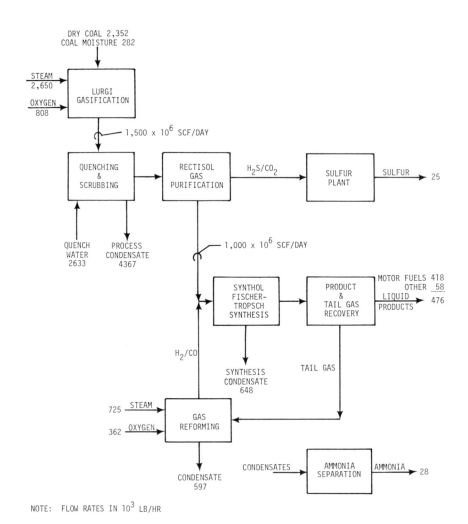

Figure 6-14
Simplified flow diagram for Sasol Fischer-Tropsch plant with a nominal output of 50,000 barrels/day of gasoline (0.65 specific gravity).

boiler feed water. The product gases are separated from the catalyst in cyclones, cooled to condense out the heavier hydrocarbon products and passed through an absorber for recovery of the lighter hydrocarbon fraction. The unconverted "tail gas" goes to a gas reforming unit, where the unwanted methane is converted back to carbon monoxide and hydrogen by partial combustion with a steam-

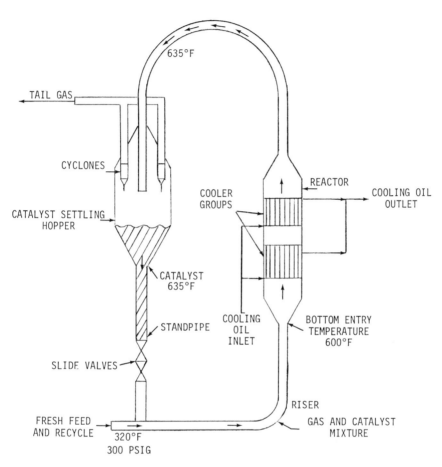

Figure 6-15
Synthol Fischer-Tropsch reactor.

oxygen mixture over a nickel catalyst. The synthesis gas is then recycled to the
Synthol reactor.

Table 6.29 is a water equivalent hydrogen balance for the new Sasol plant
designed to synthesize $1,000 \times 10^6$ scf/day of purified gas. The balance is only
approximate and is derived for the feed and product flows given in Ref. 40. The
values were estimated from mass balances using data quoted in Refs. 41 to 43.
By far the largest effluent stream is the dirty process condensate from the Lurgi
gasifiers, the quality of which has already been discussed in Section 5.5. Assuming treatment and reuse of the water within the plant, the net consumption in the

Table 6.29
Water equivalent hydrogen balance for Sasol Fischer-Tropsch plant with a nominal output of 50,000 barrels/day of gasoline (0.65 specific gravity, 20,700 Btu/lb) from 31,600 tons/day of as-received coal (10.7% moisture, 8,380 Btu/lb dry).

	10^3 lb/hr	gal/10^6 Btu*
In		
Moisture in coal fed to gasifier	282	3.44
Water equiv. of H_2 in coal	593	7.22
Steam to gasifier	2,650	32.28
Water to gas quench	2,633	32.08
Steam to reformer	725	8.83
Total	6,883	83.85
Out		
Process condensate	4,367	53.20
Water equiv. of H_2 (mainly H_2 and NH_3) in Rectisol and process condensates	111	1.35
Condensate from Fischer-Tropsch synthesis	648	7.89
Water equivalent of H_2 in hydrocarbon product	612	7.46
Condensate from reforming section	597	7.27
Water equiv. of H_2 in product NH_3	44	0.54
Total	6,379	77.71
Net water consumed in process	396	4.83

*In the liquid hydrocarbon products.

process, when measured in terms of the heating value output, is close to that for gasification, as shown in Table 5.28.

Table 6.30 is an overall thermal balance for the plant that shows the relatively low conversion efficiency when compared with hydroliquefaction processes. The cooling water requirement is correspondingly high. Assuming 25 percent of the unrecovered heat dissipated by wet cooling, the water evaporated would be 1,500 $\times$ 10^3 lb/hr or about 18 gal/10^6 Btu of gasoline.

6.8 A Summary of Water Stream Quantities

In this chapter we have examined the technologies of hydroliquefaction, pyrolysis, and Fischer-Tropsch synthesis for converting coal to clean liquid fuels. We have also looked at pyrolysis coupled with hydrotreating to convert shale to a clean synthetic crude, and solvent refining to convert coal to a clean

Table 6.30
Thermal balance for Sasol Fischer-Tropsch
plant with a nominal output of 50,000
barrels/day of gasoline (0.65 specific
gravity, 20,700 Btu/lb) from 31,600
tons/day of as-received coal (10.7%
moisture, 8,380 Btu/lb dry).

Heating Value	10^9 Btu/hr
Coal feed	19.71
Liquid products	(9.85)
Tar products*	(1.11)
Ammonia	(0.24)
Sulfur	(0.10)
Unrecovered heat	8.41
Overall conversion efficiency	57%

*56×10^3 lb/hr from Lurgi gasifier.

solid fuel. For each technology one proprietary process was presented in some detail, except for shale pyrolysis where three different ones were considered, each with a different mode of retorting. A comparison of the water streams between processes is not quite as straightforward as was the case for coal gasification to produce pipeline gas. However, the different technologies are fundamentally similar in many respects despite the outward differences. The one technology that is not directly comparable with the others is Fischer-Tropsch synthesis. Furthermore, the COED coal pyrolysis process cannot be compared directly with the other liquid fuel processes since char is a principal product in addition to liquid fuel. In a fully integrated plant the char would be gasified.

Table 6.31 summarizes for the plants examined the imported and exported water streams to produce 50,000 barrels/day of a synthetic oil or gasoline or the solid equivalent of 10,000 tons/day of solvent refined coal. We again emphasize that the results are individual examples and are not the only reasonable set of numbers. Also note that the COED water streams are not the actual streams but only 35 percent of the stream quantities. The reason for not listing the total amounts is because of the large quantity of char produced in this process in addition to the 50,000 barrel/day output of synthetic crude. The figure of 35 percent represents the fractional heating value of the oil compared with the total product heating value of char plus oil. The results are most interesting in showing that the total fresh water process requirements for all of the plants fall into a very

Table 6.31
Summary of process streams in 10^3 lb/hr for the production of 50,000 barrels/day
of synthetic oil or gasoline or 10,000 tons/day of solvent refined coal.

Process	Product	Name	In Fresh Water	Out Condensate
Synthesis	Gasoline	Fischer-Tropsch	6,008	5,612
Hydroliquefaction	Fuel Oil	Synthoil	556	388
Solvent Refining	SRC	N.M. Coal	639	604
		Wyo. Coal	715	725
Coal Pyrolysis	Syncrude	COED*	472	485
Shale Pyrolysis	Syncrude	Paraho Direct	406	522
		Paraho Indirect	465	445
		TOSCO II	666	585

*Water quantities, 35 percent of total (see text).

narrow range, except for Fischer-Tropsch synthesis. Indeed, except for Fischer-Tropsch, they are all within about 25 percent of 560×10^3 lb/hr. The low value is about 400×10^3 lb/hr for the Paraho directly heated shale pyrolysis process, and the high value is about 700×10^3 lb/hr for solvent refining a Wyoming coal. Moreover, the quantities are between about one half and one third of those found for the gasification processes (see Table 5.27), a direct result of the need for considerably less hydrogen. The very large fresh water requirement for the Sasol Fischer-Tropsch plant is a consequence of having to first gasify the coal, a procedure accomplished in the Sasol plant by using Lurgi gasifiers with a gas volume output about 4 times that for a 250×10^6 scf/day plant. The direct result of this is a fresh water requirement just about 4 times that shown for the Lurgi process in Table 5.27.

The other interesting feature of the results in Table 6.31 is how close in value the amounts of water put out as condensate are to the fresh water streams. For the most part the condensate is quite dirty. Relatively clean condensate is negligible in the shale processes and represents about 10 to 15 percent of the total condensate for all the other processes. Table 6.32 shows the net water requirements for the different processes as determined by the difference between the imported and exported streams. The largest requirement for converting coal is in the Fischer-Tropsch production of gasoline, where the hydrogen requirement is largest. This is followed next by hydroliquefaction to produce low sulfur fuel oil where the hydrogen requirement is somewhat lower. Solvent refining results in a small decrease in hydrogen content and, depending upon the coal and details of the

Table 6.32

Difference between imported and exported process waters streams for the production of 50,000 barrels/day of synthetic oil or gasoline or 10,000 tons/day of solvent refined coal.

Process	Product*	Name	Net Water Requirement	
			10^3 lb/hr	gal/10^6 Btu†
Synthesis	Gasoline	Fischer-Tropsch	396	4.83
Hydroliquefaction	Fuel Oil	Synthoil	168	1.55
Solvent Refining	SRC	N.M. Coal	35	0.31
		Wyo. Coal	(10)	(0.10)
Coal Pyrolysis	Syncrude	COED**	(13)	(0.12)
Shale Pyrolysis	Syncrude	Paraho Direct	(116)	(1.16)
		Paraho Indirect	20	0.20
		TOSCO II	281	2.81

*Product heating value 2.9–3.1 × 10^{11} Btu/day, except 2.4 × 10^{11} Btu/day for gasoline.
†In the product.
**Water quantity, 35 percent of total (see text).

process, could lead to either a small net water production or consumption, as shown in the table. Directly heated pyrolysis processes tend to be net producers of water, for example, in the Paraho directly heated process. Indirectly heated pyrolysis processes for either coal or shale may be either small net producers or consumers depending upon the grade of the raw material and the process details. This is illustrated by the TOSCO II process where more than 60 percent of the consumption shown is for water usage in wet venturi scrubbers. Except for the Fischer-Tropsch process, none of the liquefaction processes can really be considered water consumers.

Table 6.33 summarizes representative estimates of the wet cooling loads for all of the processes. The estimates of water evaporated assume that the plant is not located in an arid region. If it were, the values given could be reduced by about one half through the use of wet/dry cooling, at least in those cases where the fraction of unrecovered heat disposed of by wet cooling is greater than, say, 25 percent. The disposition of a large fraction of the unrecovered heat by wet cooling in the Synthoil process is associated with the large cooling requirements for turbines and interstage compressors. Generally, however, the fraction of unrecovered heat disposed of by wet cooling will be about the same as for gasification, the absolute range of cooling water requirements being between about 2.3 × 10^6 and 4 × 10^6 gal/day. This is not much different than that for the

Table 6.33

Water evaporated for wet cooling in the production of 50,000 barrels/day of synthetic oil or gasoline or 10,000 tons/day of solvent refined coal.

Name	Conversion Efficiency (%)	Unrecovered Heat to Wet Cooling (%)	Cooling Water Evaporated		
			10^3 lb/hr	10^6 gal/day	gal/10^6 Btu*
SRC	81	32	793	2.28	7.1
Synthoil	80	58	1,400	4.03	13.0
COED	72	25†	873	2.52	8.4
Paraho Direct	71	28	1,160	3.34	11.6
TOSCO II	68	18	850	2.45	8.5
Paraho Indirect	57	19	1,330	3.83	13.3
Fischer-Tropsch	57	25**	1,500	4.32	18.3

*In the product.
†Based on 35 percent of unrecovered heat generated in syncrude production (see text).
**Assumed (not calculated from plant design).

gasification plants. However, the heating value output of the liquid fuel plants is some 30 percent higher, with the exception of Fischer-Tropsch synthesis. In the Fischer-Tropsch case the heating value output is closely equivalent to that for a pipeline gas plant producing 250×10^6 scf/day. As before, any cooling tower blowdown water is assumed to be reused and thus does not represent a net consumption.

Chapter 6. References

1. Tetra Tech, Inc., "Energy from Coal—A State-of-the-Art Review," ERDA Report No. 76-7, U.S. Gov't Printing Office, Washington, D.C., 1976.

2. Furlong, L. E., Effron, E., Vernon, L. W., and Wilson, E. L., "The Exxon Donor Solvent Process," *Chemical Engineering Progress* **72** (8), 69–79, Aug. 1976.

3. Bodle, W. W. and Vyas, K. C., "Clean Fuels from Coal," *The Oil and Gas Journal,* pp. 73–88, Aug. 26, 1974.

4. Stiles, A. B., "Methanol, Past, Present, and Speculation on the Future," *AIChE J.* **23,** 362–375 (1977).

5. Sass, A., "Garrett's Coal Pyrolysis Process," *Chemical Engineering Progress* **70** (1), 72–73, Jan. 1974.

6. Hendrickson, T. A., *Synthetic Fuels Data Handbook.* Cameron Engineers, Inc., Denver, Colo., 1975.

7. Carlson, F. B., Yardumean, L. H., and Atwood, M. T., "The Toscoal Process for Low-Temperature Coal Pyrolysis," *Chemical Engineering Progress* **69** (3), 50, March 1973.

8. McKee, J. M. and Kunchal, S. K., "Energy and Water Requirements for an Oil Shale Plant Based on Paraho Processes," *Quarterly Colorado School of Mines* **71** (4), 49–64, Oct. 1976.

9. Freedman, S., Yavorsky, P. M., and Akhtar, S., "The Synthoil Process," *Clean Fuels from Coal Symposium II,* pp. 481–494, Institute of Gas Technology, Chicago, Ill., 1975.

10. Yavorsky, P. M., Akhtar, S., Lacey, J. J., Weintraub, M., and Reznik, A. A., "The Synthoil Process," *Chemical Engineering Progress* **71** (4), 79–80, April 1975.

11. Bureau of Mines, "Economic Analysis of Synthoil Plant Producing 50,000 Barrels per Day of Liquid Fuels from Two Coal Seams: Wyodak and Western Kentucky," Report No. ERDA 76-35, Dept. of the Interior, Morgantown, W.Va., Nov. 1975.

12. Gold, H., Goldstein, D. J., Probstein, R. F., Shen, J. S., and Yung, D., "Water Requirements for Steam-Electric Power Generation and Synthetic Fuel Plants in the Western United States," Report No. EPA-600/7-77-037, Environmental Protection Agency, Office of Energy, Minerals & Industry, Washington, D.C., May 1977.

13. Johnson, C. A., Chervenak, M. C., Johanson, E. S., and Wolk, R. H., "Scale-up Factors in the H-Coal Process," *Chemical Engineering Progress* **69** (3), 52–54, March 1973.

14. Johnson, D. A., Chervenak, M. C., Johanson, E. S., Stotler, H. H., Winter, O., and Wolk, R. H., "Present Status of the H-Coal Process," *Clean Fuels from Coal Symposium,* pp. 549–575, Institute of Gas Technology, Chicago, Ill., 1973.

15. Schmid, B. K., "Status of the SRC Project," *Chemical Engineering Progress* **71** (4), 75–78, April 1975.

16. Pastor, G. R., Keettley, D. J., and Naylor, J. D., "Operation of the SRC Pilot Plant," *Chemical Engineering Progress* **72** (8), 67–68, Aug. 1976.

17. Schmid, B. K., "The Solvent Refined Coal Process," *Symp. on Coal Gasification and Liquefaction,* Univ. of Pittsburgh, Pittsburgh, Pa., Aug. 1974.

18. Southern Services, Inc., "Status of Wilsonville Solvent Refined Coal Pilot Plant," Report No. 1234, Electric Power Research Institute, Palo Alto, Calif., May 1975.

19. Anderson, R. P., "Recycling Solvent Techniques for the SRC Process," *Chemical Engineering Progress* **71** (4), 72–74, April 1975.

20. Anderson, R. P. and Wright, C. H., "Development of a Process for Producing an Ashless, Low-Sulfur Fuel from Coal, Vol. II—Laboratory Studies, Part 3—Continuous Reactor Experiments Using Petroleum Derived Solvent," Report No. 53, Interim Report No. 8, (NTIS Catalog No. FE-449-Tl), Energy Res. & Develop. Admin., Washington, D.C., May 1975.

21. Goldstein, D. J. and Yung, D., "Water Conservation and Pollution Control in Coal Conversion Processes," Report No. EPA-600/7-77-065, Environmental Protection Agency, Research Triangle Park, N.C., June 1977.

22. U.S. Department of the Interior, "Final Environmental Statement for the Prototype Oil Shale Leasing Program," Vol. I, U.S. Gov't Printing Office, Washington, D.C., 1973.

23. McCarthy, H. E. and Cha, C. Y., "Oxy Modified In Situ Oil Process Development and Update," *Quarterly Colorado School of Mines* **71** (4), 85–100, Oct. 1976.

24. Weichman, B. E., "Oil Shale is Not Dead," *Quarterly Colorado School of Mines* **71** (4), 71–84, Oct. 1976.

25. Matzick, A., Dannenberg, R. O., Ruark, J. R., Phillips, J. E., Lankford, J. D., and Guthrie, B., "Development of the Bureau of Mines Gas-Combustion Oil-Shale Retorting Process," Bulletin No. 635, Bureau of Mines, Dept. of the Interior, Washington, D.C., 1966.

26. Ruark, J. R., Sohns, H. W., and Carpenter, H. C., "Gas Combustion Retorting of Oil Shale Under Anvil Points Lease Agreement: Stage II," Report of Investigations No. 7540, Bureau of Mines, Dept. of the Interior, Washington, D.C., 1971.

27. Jones, J. B., Jr., "Paraho Oil Shale Retort," *Quarterly Colorado School of Mines* **71** (4), 39–48, Oct. 1976.

28. Schora, F. C., Jr., Feldkirchner, H. L., Tarman, P. B., and Weil, S. A., "Shale Gasification Under Study," *Hydrocarbon Processing,* 89–91, April 1974.

29. Weil, S. A., Feldkirchner, H. L., and Tarman, P. B., "Hydrogasification of Oil Shale," *Shale, Oil, Tar Sands and Related Fuel Sources* (T. F. Yen, ed.), pp. 55–76, Advances in Chemistry Series 151, American Chemical Society, Washington, D.C., 1976.

30. Colony Development Operation, "An Environmental Impact Analysis for a Shale Oil Complex at Parachute Creek, Colorado, Part I—Plant Complex and Service Corridor," (also corrected water system flow diagram, personal communication), Atlantic Richfield Co., Denver, Colorado, 1974.

31. Schmalfeld, P., "The Use of the Lurgi-Ruhrgas Process for the Distillation of Oil Shale," *Quarterly Colorado School of Mines* **70** (3), 129–145, July 1975.

32. Whitcombe, J. A. and Vawter, R. G., "The Tosco-II Oil Shale Process," Paper No. 40a, AIChE 79th National Meeting, March 1975.

33. Cook, E. W., "Organic Acids in Process Water from Green River Oil Shale," *Chemistry and Industry,* p. 485, May 1, 1971.

34. Hubbard, A. B., "Method for Reclaiming Waste Water from Oil-Shale Processing," American Chemical Society, Div. of Fuel Chemistry Preprints **15,** No. 1, pp. 21–25, March–April 1971.

35. Jones, J. F., "Project COED (Char-Oil-Energy-Development)," *Clean Fuels from Coal Symposium,* pp. 323–342, Institute of Gas Technology, Chicago, Ill., 1973.

36. Scotti, L. J. *et al.,* "Char Oil Energy Development," R&D Report No. 73—Interim Report No. 2, Period of Operation July 1972–June 1973, Office of Coal Research, Department of the Interior, Washington, D.C., 1974.

37. Kalfadelis, C. D. and Magee, E. M., "Evaluation of Pollution Control in Fossil Fuel Conversion Processes, Liquefaction: Section 1. COED Process," Report No. EPA-650/2-74-009-e, Environmental Protection Agency, Office of Research & Development, Washington, D.C., Jan. 1975.

38. Hamshar, J. A., Terzian, H. D., and Scotti, L. J., "Clean Fuels from Coal by the COED Process," *Symp. Proc.: Environmental Aspects of Fuel Conversion Technology,* pp. 147–157, Report No. EPA 650/2-74-118, Environmental Protection Agency, Research Triangle Park, N.C., Oct. 1974.

39. O'Hara, J. B. *et al.*, "Fischer-Tropsch Complex Conceptual Design/Economic Analysis, Oil and SNG Production," R&D Report No. 114—Interim Report No. 3, Report No. FE-1775-7, Energy Res. & Develop. Admin., Washington, D.C., Jan. 1977.

40. Hoogendoorn, J. C., "New Applications of the Fischer-Tropsch Process," *Clean Fuels from Coal Symposium II,* pp. 343–358, Institute of Gas Technology, Chicago, Ill., 1975.

41. Hoogendoorn, J. C., "Experience with Fischer-Tropsch Synthesis at SASOL," *Clean Fuels from Coal Symposium,* pp. 353–365, Institute of Gas Technology, Chicago, Ill., 1973.

42. Hoogendoorn, J. C., "Gas from Coal with Lurgi Gasification at SASOL," *Clean Fuels from Coal Symposium,* pp. 111–125, Institute of Gas Technology, Chicago, Ill., 1973.

43. Hoogendoorn, J. C. and Salomon, J. H., "SASOL: World's Largest Oil-from-Coal Plant," *British Chemical Engineering* **2,** 238–244, 308–312, 368–373, 418–419; May, June, July, Aug., 1957.

7
Mining, Disposal, and Other Water Uses

7.1 Mining and Fuel Preparation

In the preceding chapters we examined the process and cooling water requirements for synthetic fuel plants. In any mine-plant complex, however, water is also needed to mine and prepare the coal or shale, and to dispose of the residual coal ash and spent shale. Water may also be needed for reclamation of surface mined land, for sulfur disposal, and for a variety of other smaller mine and plant requirements. As a rule, many of the mine and disposal water needs can be met with lower quality waters.

The water requirements for mining are generally not strongly affected by the conversion process, except as it determines the actual quantity of material to be mined. However, the mine location and whether the mining is surface or underground are strong determinants of the quantity of water consumed.

Coal Mining

For mine-plant complexes where the coal is to be converted near the mine, coal washing is assumed not to be necessary. Depending on the coal, the mine, and detailed economic factors, this need not always be the case. The purpose of washing coal is to reduce the ash and sulfur levels, but all of the conversion processes incorporate sulfur removal and recovery. Moreover, the processes themselves almost always utilize dry coal, so there would be no advantage to removing the ash at the expense of adding moisture. The coal preparation is therefore taken to consist of breaking, dry screening, and crushing operations as required. Under these conditions, the material yield is typically 95 percent of what is mined and the heating value recovery is 99 percent.[1] Thus the rate of coal mined is taken to be equal to the rate at which coal is utilized in the conversion process. An exception to this is the Lurgi process described in Section 5.3, which cannot accept fines. In that process the coal mining rate is about 15 to 20 percent more than the utilization rate.

For a fixed size plant, as measured by the calorific value of the product fuel, and for a given conversion efficiency, the rate of coal mined is determined by its heating value. Average heating values for the three major coal ranks are: bituminous 13,000 Btu/lb, subbituminous 9,800 Btu/lb, and lignite 6,800 Btu/lb.[2] The range of heating values is about a factor of two. This provides a measure of the variation in coal mining rates to be expected for different coals. For three different low sulfur, western coals, Table 7.1 shows the amount of coal that must be mined daily to produce pipeline gas, synthetic crude, and solvent refined coal for

Table 7.1
Coal mining rates for standard size plants producing synthetic fuels
from western coals.*

Process	Output	Location	Coal Mined (tons/day)	As-Received HHV (Btu/lb)
Synthane	250×10^6 scf/day	Wyo.	23,000	8,905
SRC	10,000 tons/day	N.M.	23,400	9,613
Synthoil	50,000 barrels/day	N.M.	25,200	8,310

*Taken from examples given for these processes in Chapters 5 and 6.

the nominal outputs of standard size plants. The amounts of coal required by all three processes for the plant sizes considered are seen to be about the same.

In estimating water needs for coal mining, the distinction must be made between surface and underground mining. Whether the mine is located in the humid eastern and central parts of the country or in the arid and semi-arid western regions will also make a difference when estimating. For the purpose of controlling dust when surface mining coal, water is generally sprayed on the unpaved haul roads, mine benches, and overburden placement areas. The problem of fugitive dust control is greatest in the dry arid and semi-arid western regions. In these regions, it is assumed that the haul roads and mine area can be kept in a wetted condition through an annual deposition of water equal to the net annual evaporation rate.[3] Any rainfall is taken to be an additional safety factor and is not subtracted from the amount laid down, because how much is absorbed and how much runs off are variables. The rate at which water is laid down is therefore given by the product of the evaporation rate and the area on which the dust control is practiced. In Table 7.2 the average annual pond evaporation rates are shown for the coal mining regions in the western United States.

Table 7.2
Annual pond evaporation rates in western
coal mining regions.

State	Evaporation Rate (in/yr)
New Mexico	61
Wyoming	54
Montana	49
Colorado	45
North Dakota	45

The extent of the haul roads for a surface mine will depend mostly upon the active mine area.[4] In one simple mine model, for example, the active mine area is the area that is mined annually.[3] This area is equal to the annual mining rate divided by the coal yield per unit of land area stripped. As Table 7.1 shows, the amount of coal required for all of the standard size plants is quite close to 1,000 tons/hr or about 8×10^6 tons/yr assuming 8,000 hrs/yr of plant operation (about 90 percent load factor). Table 7.3 gives the average coal yields per acre for the western coal regions and the corresponding active mine areas for 8×10^6 tons/yr of production.

Mine haul roads constitute about 10 percent of the active mine area,[4] a figure consistent with estimates for model western surface mines.[3] An additional 2 percent of the total mine area is estimated to be wet down at the mine bench and overburden placement areas.[3] With the total area to be wet down equal to about 12 percent of the mine area, and with the annual rate at which water is to be laid down equal to the annual evaporation rate, it is a simple matter to find the water consumption as the product of the area to be wetted down and the evaporation rate. (Consistent units and 8,000 hrs/yr operation are assumed when converting from yearly to hourly rates.) The results are shown in Table 7.4 and to within the accuracy of the estimates may be used for any of the standard size plants considered in the text. For larger or smaller sizes they may be scaled up or down in correspondence to the output. We strongly emphasize that these results are for one surface mine model. Others are possible with differing results, depending upon the detailed nature of the mine layout and operation.

It should also be mentioned that the application of calcium chloride to the mine roads when they are wet is effective in controlling dust[4] and can reduce the amount of water needed.

Table 7.3
Coal yields per acre in western coal mining regions and estimated active surface mine areas for 8×10^6 tons/yr production.

State	Coal Yield (tons/acre)	Mine Area (acres)
Wyoming	90,000	90
Montana	40,000	200
New Mexico	37,000	215
North Dakota	25,000	320
Colorado	23,000	350

Table 7.4
Water consumption in mine and road dust control for
8×10^6 ton/yr western surface coal mines.

State	gal/min	lb water/10^3 lb coal
Wyoming	33	8
Montana	67	17
New Mexico	89	22
North Dakota	98	25
Colorado	107	27

In the Appalachian and Illinois Basins, the mean annual precipitation rates of about 35 to 45 in/yr exceed the evaporation rates. Therefore, the method of estimating the surface mine dust control water requirements for the western regions would be quite inappropriate for the East. In the West, the evaporation rates range from 45 to 60 in/yr in comparison with annual precipitation rates of 8 to 15 in/yr. Although keeping the roads wet is the most common method for holding down dust in eastern surface mines, the amount is not large and will be neglected here.

Most underground mining of coal is in the Appalachian Basin, where water supply is generally not a critical factor. The main need for water in an underground mine is to suppress airborne respirable dust. In conventional underground mining, a cutting machine cuts a slice under the coal seam, following which blastholes are drilled. The coal is fragmented by blasting and then transported to the surface. All of these operations generate dust, though cutting at the mine face is the principal source. Water spraying and wet scrubbing of dust-laden air is used extensively to capture the dust. The net water consumption for dust control in underground coal mining varies considerably (for examples, see Ref. 5). Precise figures are difficult to give, since they depend critically upon the methods of dust control and the extent of water recovery practiced. From data cited in Ref. 4 for mines where water conservation would most likely be practiced, Table 7.5

Table 7.5
Water consumption in dust control for 6×10^6 ton/yr
underground coal mines.

Probable Range	gal/min	lb water/10^3 lb coal
Low	100	33
High	300	100

presents a probable range of consumption for a 6×10^6 ton/yr underground mine. A mine with this output of a high heating value eastern coal would meet the coal needs of the standard size synthetic fuel plants considered.

Shale Mining

Although oil shale can be surface mined, the amount of shale that can be economically recovered by this method is relatively small, and therefore only the water requirements for underground mining will be presented. The underground mining of shale is similar to that for coal. Estimates of the water consumption for underground shale mining have been reported for the Paraho and TOSCO II integrated oil shale mine-plant designs.[6,7] For the Paraho designs, however, only the combined requirements for mining and crushing are given.[7] On the basis of the split between the requirements as given in the TOSCO II design, we have arbitrarily assumed that 70 percent of the dust control water is for the mine and the remaining 30 percent is for the crushing and other dust control operations.

Table 7.6 summarizes the water consumed for dust control in underground mining of shale for the three integrated plant designs examined in Section 6.5. The mining rates are four to six times those for an underground eastern coal mine. Part of the difference in the mining rates between the Paraho and the TOSCO II processes is a consequence of the difference in the grade of shale assumed to be mined. It should also be mentioned that because the Paraho retort (see Section 6.4) cannot accept fines, about 5 percent more shale must be mined than can be used.[7] The reported water needs show about a 30 percent difference in the unit water requirements between the two designs, although the absolute amounts are quite close. Better accuracy is probably not possible at this stage.

Table 7.6
Water consumed in dust control for underground shale mines integrated with shale oil plants producing 50,000 barrels/day of synthetic crude.

Process	Shale Grade gal/ton	Shale Mined tons/day	Water Consumed	
			gal/min	lb water/10^3 lb shale
TOSCO II	35	73,300	389	32
Paraho Direct	30	92,000*	354†	23
Paraho Indirect	30	105,000*	404†	23

*5 percent more than used.
†Based on 70 percent to mining, 30 percent to crushing.

The TOSCO II value of 32 lbs water/10^3 lb shale provided the basis for the lower estimate given in Table 7.5 for coal mining, where gas removal requirements and greater explosion hazards would be expected to require a greater degree of dust control than would shale mining.

Fuel Preparation

In the preparation and handling of the coal or shale for delivery to the conversion plant, dust is generated in the stages of loading and unloading, breaking, conveying, crushing, general screening, and storage. Water is required to hold down this dust. Water could also be required for coal washing, but it is assumed that this will generally not be needed. Therefore, dust control is the principal water need for fuel preparation.

Many of the ways of preventing dust from becoming airborne are similar to those used in controlling mine dust and include the application of water sprays and non-toxic chemicals, the use of dry or wet dust collectors, and the use of either partial or total enclosure. We shall assume that the principal dust-generating sources will be enclosed and that where feasible the air will be circulated and dry bag dust collection employed. Whenever coal pulverization is necessary we consider, as is normally the case, that this will be done under conditions of total enclosure with no fugitive dust or hold-down water requirements. In inactive storage the use of water for holding down dust can be minimized by the use of non-toxic chemicals.

Despite design precautions to prevent dust from becoming airborne, in large scale plants with many transfer points, transfer belts, surge bins, storage silos, and active storage sites, water sprays must be employed to wet down the coal or

Table 7.7
Water consumed in fuel handling and dust control.

Process	Coal/Shale Mined tons/day	Water Consumed gal/min	lb water/10^3 lb fuel
Eastern Coal Conversion	18,000	30–45	10–15
Western Coal Conversion	24,000	45–60	10–15
TOSCO II	73,000	167	14
Paraho Direct	92,000	152	10
Paraho Indirect	105,000	173	10

shale. This is also generally necessary with breaking and primary crushing operations. Table 7.7 gives the oil shale water requirements as scaled from the designs of Refs. 6 and 7, along with an estimated range for crushing and handling of eastern and western coals in the nominal amounts used for the plant sizes considered.

7.2 Ash Disposal, Flue Gas Scrubbing, and Reclamation

Ash enters a conversion plant in the coal and if the coal is not washed will leave from a furnace or gasifier as fly ash, bottom ash, or slag. Fly ash refers to the particulates of ash entrained in the exiting flue gases. Bottom ash or slag is that part of the ash that is quenched and removed from the bottom of the furnace or gasifier. Water is needed for the collection and disposal of the different ashes. An exception is when fly ash is removed dry and water is needed only for disposal. In industrial practice the amount of water consumed in ash removal by different handling schemes varies widely. The estimates given here assume the avoidance of wasteful consumption but do not represent minimum requirements. Nor are they the only reasonable ones.

The rate of ash disposal is set by the amount of ash in the coal and the rate at which coal is fed to the plant. For a feed of 24,000 tons/day of a 10 percent ash coal, which is representative for a standard size conversion plant using a western coal, this amounts to 2,400 tons/day of ash to be collected, handled, and disposed. The amount of water consumed in the disposal is principally dependent on the handling scheme. It is also somewhat dependent on the process and integrated plant design, to the extent they affect the fraction of ash removed as fly ash and the fraction removed as bottom ash or slag. For bottom ash and fly ash, the amount of water consumed in disposal is quite different. Two other characteristics of the process affect the amount of water evaporated in bottom ash or slag removal: the ash or slag temperature and the amount of heat radiated from the reactor into the bottom ash removal hopper.

In most gasifiers the ash leaves as bottom ash or slag. This includes the Hygas, Lurgi, and Koppers-Totzek gasifiers discussed in previous chapters. In the Koppers-Totzek gasifier, some of the ash is removed from the gas, but it is treated as bottom ash. However, in the integrated plant design some coal may be burned separately in boilers to generate steam. This coal fraction generates both fly ash and bottom ash. For modern, dry bottom, pulverized coal furnaces about 20 percent of the ash leaves as bottom ash and 80 percent as fly ash. Wet bottom and cyclone furnaces develop slag in place of bottom ash and have lower fly ash fractions. Present trends, however, are away from the use of cyclone furnaces, because the high combustion temperatures result in high emissions of oxides of nitrogen, and away from wet bottom furnaces because of the maintenance problems associated with slag removal. Consistent with these trends, dry bottom boilers are assumed to be used when coal is burned to generate steam.

In the Hygas integrated pipeline gas plant discussed in Section 5.3, 14 percent of the coal fed to the plant is burned in boilers to generate steam, resulting in about 10 percent less bottom ash compared to an estimate that assumed all the coal was fed to the gasifier. One design proposed for a Lurgi pipeline gas plant in New Mexico[8] uses about the same fraction of the feed coal to generate steam in boilers as the Hygas design does. In the Synthane gasifier all of the ash in the coal comes out in the char. For the integrated plant design discussed in Section 5.2, the char is fed to boilers to generate steam. These boilers are assumed to be similar to the pulverized coal, dry bottom furnaces with 20 percent of the ash leaving as bottom ash and 80 percent as fly ash. In the CO_2 Acceptor process, part of the ash is separated out from the acceptor bed; the part produced by the burning of char in the regenerator is separated out by cyclones from the regenerator off-gas. All of this ash may be considered to be handled for disposal like bottom ash. In solid and liquid fuel production by the Solvent Refined Coal, Synthoil, and Fischer-Tropsch processes, all of the ash is assumed to come out as bottom ash from the gasifiers in the plants. The COED coal pyrolysis process produces large amounts of char containing most of the coal ash, but in a fully integrated plant this char would be gasified and the ash generally removed as bottom ash.

To summarize, except for the Synthane and Hygas designs, the ash from all of the other coal conversion plants discussed may be considered to be handled as bottom ash. The Hygas plant generates about 90 percent bottom ash and 10 percent fly ash, and the Synthane plant generates 20 percent bottom ash and 80 percent fly ash. Other gasifier plant designs not considered here, where coal is used to generate steam, may be assumed to have a distribution similar to the Hygas example.

Bottom Ash Water

Bottom ash and slag handling water requirements for gasifiers may be drawn from related experience with boilers. Bottom ash and slag are usually handled by sluicing with water to settling ponds or dewatering bins. The effluent from the bins may be either discharged or recycled. Here we assume the overflow or clarified water is recycled, consistent with reasonable water conservation practice. The bottom ash from a dry bottom boiler is collected and quenched in hoppers located beneath the boiler but not structurally attached to it. The hoppers are sluiced periodically and then refilled. The hoppers are provided with water seals which require a continuous flow of water that leaves at a temperature typically in the range of 150° to 170°F. Slag tanks below wet bottom and cyclone boilers may be sluiced continuously to prevent blockage. In either bottom ash or slag removal, there are two main sources of water consumption: evaporation from the hopper or tank below the furnace; and occluded sluice water in the settled ash, which cannot easily be recovered.

An estimate of the water evaporated is difficult to make on purely theoretical grounds, principally because the cooling and seal water must not only cool down the ash but also must absorb the radiative heat input from the bottom of the boiler (or gasifier). In modern large scale furnaces, the radiative heat input may be two or more times the heat input from the hot ash or slag. If ash at 2,000°F with a specific heat of about 0.2 Btu/lb °F were cooled to 150°F by evaporating water, 0.37 lb water/lb ash would be needed. By itself this would be an overestimate of the ash quench water evaporated. However, there is also a large radiative heat input to the quench water from the boiler. Large scale boiler practice shows a typical range for evaporated water of 0.3 to 0.6 lb water/lb ash. For 2,400 tons/day of ash this amounts to 120 to 240 gal/min of water evaporated (60×10^3 to 120×10^3 lb/hr). This range may be scaled either up or down in direct correspondence to the bottom ash removal rate.

The underflow from bottom ash dewatering bins has the consistency of wet concrete and contains about 35 weight percent water. If the ash is settled in ponds, the settled ash has about a 50 weight percent moisture content. Assuming that the sluice water not occluded in the ash will be recycled, this means that the consumption of sluice water will range between about 0.5 and 1 lb water/lb ash, depending upon the dewatering method used. For 2,400 tons/day of ash, this amounts to 200 to 400 gal/min (100×10^3 to 200×10^3 lb/hr). Recycling the sluice water tends to concentrate the dissolved solids; if the water is alkaline, calcium carbonate as well as calcium sulfate will precipitate and deposit as scale. This may not be a particular problem in ash hoppers or slag tanks. However, when buildup of dissolved solids is a problem, it can be limited by blowdown or sidestream treatment,[9] similar to the water recycling methods in a cooling tower.

Fly Ash Water

Fly ash can be collected and handled in a number of ways with different amounts of water consumed. The two most common means are dry collection by electrostatic precipitation and wet collection by scrubbers. If the fly ash is collected wet, it must also be handled wet. If it is collected dry it can be handled either wet or dry. When land and water are both cheap and readily available near the plant, then wet handling may be the preferred method. Despite this, however, effluent limitation guidelines on fly ash transport water may lead to the use of dry fly ash handling in new plants. Considering this possibility and reasonable water conservation practice, we shall assume that the fly ash is both collected and handled dry. The fly ash is therefore taken to be recovered from the flue gases using dry electrostatic precipitators ahead of the flue gas desulfurization scrubbing.

Dry fly ash is transported by pneumatic conveying from the precipitators to storage silos. It may then be disposed or used in a number of ways. In strip mining, fly ash will probably be used for landfill and reclamation. The fly ash is generally alkaline and can thus substitute for limestone to neutralize the strip mine spoils. It also enhances plant survival and growth, improves the quality of the spoil, and establishes a good cover.[4] In underground mining the fly ash will likely be returned to the mine. In either surface or underground mining, water is usually added to the fly ash by spraying or mixing to prevent dusting during unloading from silos and transportation to the disposal or utilization site. The amount of water added for dust control may vary somewhat, but about 10 weight percent is a representative figure. If too much water is added the fly ash could set up and harden. If the fly ash is for landfill, water is added up to 20 to 30 weight percent at which value fly ash has a maximum compaction density and strength.[10] For our purposes water used for dust control will not be distinguished from water used for landfill; instead an average 20 weight percent of water is assumed to be added. This is equivalent to a water addition rate of 0.25 lb water/lb ash. (This water need not be a high quality water and may be blowdown from a cooling tower or from a bottom ash sluicing system.) For the Synthane plant, with 80 percent of the ash removed as fly ash, the water consumption amounts to about 80 gal/min (40×10^3 lb/hr) for a total ash disposal rate of 2,400 tons/day. For the Hygas plant, the water consumed would be about 15 percent of this.

Flue Gas Desulfurization Water

Flue gas desulfurization takes place following the fly ash removal. As noted in

Section 3.7, we assume that sulfur dioxide is removed by a lime/limestone or sodium-lime wet scrubbing process. A general expression (Eq. 3.26) was derived for the quantity of water leaving in the scrubbed flue gas, based on the assumption that the flue gas is saturated with water vapor at the temperature and pressure at which it leaves the final absorber before any reheating. An estimate of the slurry water requirement for the disposal of the spent scrubber sludge as a function of the coal sulfur content and slurry concentration was given by Eq. (3.27).

Equations (3.26) and (3.27) have been applied to a char and a number of coals with compositions shown in Table 7.8. An important characteristic of the char is that it is moisture free. The corresponding makeup water requirements are shown in Table 7.9. To calculate the water remaining in the scrubber sludge, the solids concentration is taken to be 40 weight percent, which represents a well dewatered sludge but one that has not been vacuum filtered or centrifuged. If the scrubber sludge is to be disposed of as landfill, for physical stability it may be necessary to dewater it to as high as 65 to 70 percent solids. This dewatering is greatly facilitated by adding dry fly ash.[11]

From the results in Table 7.9, the largest single factor in the flue gas water requirement is seen to be the moisture content of the fuel. On the other hand, the slurry water makeup is directly proportional to the coal or char sulfur content. The approximate range of makeup water is between 0.3 and 1.2 lb/lb coal or char. For the Hygas example, where the coal feed to the boilers is 2,550 tons/day of subbituminous coal, the water requirement is 240 gal/min (120×10^3 lb/hr). For the Synthane example the char feed to the boilers is 4,030 tons/day, repre-

Table 7.8
Representative coal and char compositions for estimating wet flue gas desulfurization makeup water (weight fraction).

	Carbon	Hydrogen	Oxygen	Sulfur	Moisture
Coal					
Wet lignite	0.45	0.03	0.15	0.006	0.3
Med. wet lignite	0.51	0.034	0.17	0.007	0.2
Dried lignite	0.58	0.039	0.19	0.008	0.1
Subbituminous	0.54	0.04	0.11	0.009	0.1
Bituminous	0.70	0.05	0.07	0.04	0.15
Bituminous	0.70	0.05	0.07	0.03	0.05
Bituminous	0.70	0.05	0.07	0.02	0.05
Char*	0.64	0.01	0.014	0.003	0

*From Synthane example in Section 5.2.

Table 7.9
Flue gas and scrubber sludge makeup water for wet flue gas desulfurization
(lb water/lb fuel).

	Flue Gas	Scrubber Sludge	Total
Coal			
Wet lignite	0.21	0.05	0.26
Med. wet lignite	0.38	0.06	0.44
Dried lignite	0.55	0.07	0.62
Subbituminous	0.49	0.08	0.57
Bituminous	0.82	0.35	1.17
Bituminous	0.81	0.26	1.07
Bituminous	0.80	0.17	0.97
Char	0.70	0.02	0.72

senting about 20 percent of the heat input of the coal feed. The makeup water requirement is about 480 gal/min (240 × 10³ lb/hr) or just twice that for the Hygas example. Without specifying the amount of coal or char burned in the boilers or the moisture and sulfur contents of the fuel, simple generalizations cannot be made on the quantity of makeup water required for flue gas desulfurization.

Reclamation Water

Reclaiming strip mined land in arid and semi-arid regions may require supplemental irrigation water to establish a soil stabilizing plant cover on mine spoils. Coal mined areas with greater than 10 inches of mean annual precipitation can be reclaimed without supplemental irrigation according to Ref. 12. Where there is less than 10 inches of annual rainfall, partially reshaped coal mine spoils can be revegetated with supplemental irrigation of about 10 inches during the first growing season, and no further requirement during subsequent growing seasons.[13] In all the coal mining regions of the United States, the only major area where there is less than 10 inches of annual rainfall is in New Mexico. For a coal yield of 37,000 tons/acre and an 8 × 10⁶ ton/year mine, the water requirements are about 30 lb/10³ lb coal or 120 gal/min (60 × 10³ lb/hr).

Of course more than water is needed to reclaim the land. Proper reclamation techniques must be employed including returning topsoil, slope reduction, grass drilling, timely seeding, fertilizer addition, and mulching.[4]

7.3 Spent Shale Disposal and Revegetation

Few topics in the development of a synthetic fuels industry in the United States have generated as much controversy as the disposal of spent shale from retorting operations. The questions raised have concerned the difficulties in disposing of the large quantities of processed shale, the large amounts of water needed for the disposal and revegetation of the spent shale piles, and the possibilities of environmental damage from leaching of the disposed shale either from natural precipitation or from the dirty process waters that may be used for disposal. These points shall be addressed with the emphasis on the water management aspects.

Approximately 80 to 85 percent of high grade raw shale remains as spent material after retorting. If the oil shale grade is specified, the fraction of the raw shale to be disposed may be estimated by Eq. (3.3), assuming the inorganic matter calculable from the equation to represent the spent shale. Table 7.10 lists the quantities mined, retorted, and disposed for the integrated plants considered in Section 6.5. These values have been scaled from the values in the reported designs listed in Table 6.18 to an integrated plant output of 50,000 barrels/day of synthetic crude. An oil shale industry producing 10^6 barrels/day of synthetic crude would need to mine 1.5×10^6 to 2×10^6 tons/day of shale. Large as this may seem, the Kennecott Copper Co. at its strip mine near Salt Lake City, Utah, mined more than 0.75×10^6 tons/day during the period in which it enlarged its operations to their current capacity (Ref. 14, p. 271).

From each plant, between 60,000 and 90,000 tons/day of spent shale must be disposed. The processed shale from the TOSCO II retorting is a fine, black, sandy material with about 60 percent of the particles passing through a 200 mesh screen (.003 inch opening) and about 35 percent passing through a 325 mesh screen (.0018 inch opening).[6] The total carbon content is about 5 weight percent. The processed shale from the directly heated Paraho retort consists of lumps

Table 7.10
Oil shale quantities in tons/day for integrated plants producing 50,000 barrels/day of synthetic crude.

Process	Grade (gal/ton)	Mined	Fines	Spent Shale	Disposal
TOSCO II	35	73,000	—	60,000	60,000
Paraho Direct	30	92,000	4,000	71,000	75,000
Paraho Indirect	30	105,000	5,000	85,000	90,000

about $\frac{3}{8}$ to $1\frac{1}{2}$ inch in size[15] containing about 1.5 percent residual carbon. The shale from the indirectly heated Paraho retort may be expected to be about the same size, although its residual carbon content is in the range of 3 to 5 percent.

Different procedures with considerably different water needs have been proposed for the disposal of the TOSCO and Paraho spent shales. In the TOSCO II design shown in Fig. 7-1, the spent shale leaving the cooler is moisturized to approximately 15 percent moisture content in a rotating drum moisturizer. Steam and processed shale dust produced in the moisturizing procedure are passed through a venturi wet scrubber to remove the dust before discharge to the atmosphere. The moisturized spent shale is transported by a covered conveyor belt to the disposal area and then spread and compacted to a density of about 90 pounds of dry spent shale per cubic foot. During the transport, spreading, and compac-

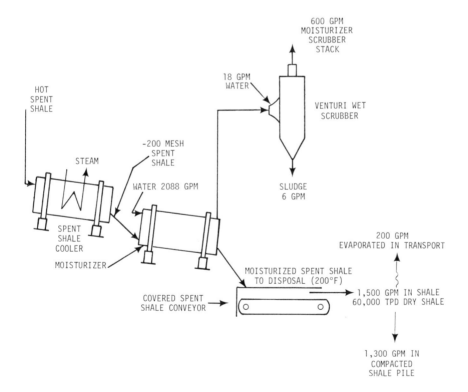

Figure 7-1
TOSCO II spent shale disposal process with quantities appropriate to an integrated plant producing 50,000 barrels/day of synthetic crude.

tion operations, about 13 percent of the added moisture evaporates. This leaves about a 13 percent in-place moisture content, defined as an optimum for compaction and setting purposes.[16]

The importance of the moisturizing is that the addition of the water to the TOSCO II type processed shale, at a predetermined shale temperature, leads to cementation of the shale after compaction. This cemented shale appears to permanently "freeze in" the moisture that was added,[6] much of which was dirty process water. Moreover, the shale becomes effectively impermeable and resists percolation so that soluble salts cannot be leached out.[16] Tests show that to percolate water through a 4.5 ft thick test bed of compacted shale requires the equivalent of 7 in/day of rain for seven days[16] (also see Ref. 14, p. 239). Processed shale piles in a TOSCO II commercial embankment are designed for a maximum depth of 700 to 800 ft and an average depth of about 250 ft.

In the TOSCO II design of Ref. 6 the spent shale is to be disposed of in a canyon. The shale is compacted into a shallow embankment and benched to decrease erosion. A flood control reservoir is located above the canyon to divert water from the canyon. Any runoff from the embankment is diverted back to the plant for use as moisturizer water.

After 20 years of operation of a 50,000 barrel/day plant the compacted spent shale would cover an area of approximately 800 acres.[6] This is an average of about 40 acres/yr and for a compaction density of 90 lbs/ft^3 would correspond to a mean height of 250 ft. Irrigated revegetation will be undertaken as permanent surfaces are created by the fill. Prior to revegetation, water spraying will be used to control dust.

In the Paraho design concept for spent shale disposal,[15] an "earth" dam constructed of retorted shale would be built at the mouth of a valley selected for a disposal area. The valley itself would be lined with a heavy compacted, impervious layer of retorted shale. By adding about 20 weight percent water prior to compaction, the shale cements up and the shale layer would thus be made impermeable. The valley would then form a lined basin ("bath tub") into which the retorted shale could be deposited. It is assumed that any precipitation leaching through the spent shale would be held within the basin. The important point here is that the spent shale would be compacted but not be wetted down, except for controlling dust and for revegetation. Tests have shown a compaction density of about 90 lbs/ft^3 can be obtained, which is similar to that gotten for spent shale that has been wetted down. It is estimated that less than one percent of the total volume of the shale disposed would have to be wetted to obtain a material of high strength and low permeability. Such a disposal scheme would substantially reduce the water requirements for oil shale plants. On the other hand, the TOSCO procedure, although more water consuming, has had sufficient long-term testing for there to be reasonable assurance that serious environmental problems will not be encountered.

Revegetation of spent oil shale piles has been studied for some time.[6,17] Because of differences in the vegetative patterns where oil shale is mined and differences in the nature of the processed shale, no simple estimate can be given on the amount of water required to establish vegetative cover (Ref. 14, pp. 253–265). Among the difficulties in revegetating processed shale are its high salt content and lack of nutrients. In the case of TOSCO spent shale, its black color can result in very high surface temperatures during the summer months making germination difficult. For the Paraho processes, a much higher water requirement for revegetating the spent shale from the indirect retort than from the direct retort has been ascribed to the higher residual carbon in the former.[7]

Estimates of the water needed to revegetate and to control dust prior to revegetation must rely solely on results of tests on the specific processed shale in the particular disposal area. In any case, the amount of water required will be relatively large compared, for example, to reclaiming strip-mined coal lands in an arid region. At least 4 ft of water are required for leaching the salt from the spoils. Additionally, two to three times this amount could be required over, say, a five year period to ensure a successful cover. To some extent, the amount of water needed for dust control will depend on how rapidly a vegetative cover is established.

Table 7.11 summarizes the reported data on the water requirements for spent shale disposal for the three designs examined in Chapter 6. The Paraho requirements as reported did not distinguish between that water needed for dust control

Table 7.11
Water requirements for spent shale disposal from integrated plants producing 50,000 barrels/day of synthetic crude.

Process	Moisturizing gal/min	Dust Control & Revegetation gal/min	Total gal/min	Total lb water per 10^3 lb spent shale
TOSCO II	2,106	672*	2,778	278
Paraho Indirect	—	2,320	2,321	155
Paraho Direct	—	886	886	71

*Dust control 278 gal/min. Revegetation water of 394 gal/min is 20 year average.

and that for revegetation. The estimate for the revegetation water for the TOSCO II spent shale piles was derived from averaging 78 gal/min for years 1 to 11 of the plant and 780 gal/min for years 12 to 20. The reader should be reminded again that these figures have been scaled upward somewhat from the values quoted for the plant size in Ref. 6. Despite the fact that the processed shale from the indirectly heated Paraho retort is disposed of dry, the total disposal water requirement is more than half that for the TOSCO II process as measured per unit weight of spent shale. Again, the spent shales from these two indirect processes have similar residual carbon contents. The very low water requirement for the directly heated Paraho retort, at least compared to the other two processes, is evident. Per unit weight of spent shale disposed, the water requirement for this process is one fourth that for the TOSCO II process. In the same terms, the two processes have almost identical dust control and revegetation requirements. The value for the TOSCO process is 67 lb water/10^3 lb spent shale compared to 71 lb water/10^3 lb spent shale for the Paraho direct retort.

The information presented in this section on spent shale disposal demonstrates that the technical problems in general and water management problems in particular are understood well enough to justify the expectation that any problems that might arise in a full scale plant could be controlled.

7.4 Other Mine-Plant Water Needs

There are within an integrated mine-plant synthetic fuel complex a number of consumptive uses of water, mostly of medium quality, other than those already considered. Even when taken together, these uses are generally quite small in comparison with the process or cooling water requirements. However, in any water balance for the complex, the water consumed in these uses should still be taken into account, among these the sanitary, potable, service, and fire water needs in both the plant and the mine. The quantities of water consumed for these purposes are related to the number of people engaged in the mine and plant activities. Also related to the integrated plant population are the water needs for any satellite town.

Another use not previously considered is water for fugitive dust control within the plant itself. This usage is governed for the most part by the coal or shale throughput. Finally, there is the loss due to evaporation from on-site reservoirs used for water storage or settling basins used for clarification of source water.

The number of people employed in the mine will primarily depend on whether it is a surface or underground mine and on the mine output, and the number of

Table 7.12
Representative ranges of mine and plant personnel for 250 × 10⁶ scf/day or
50,000 barrel/day integrated synthetic fuel plants.

Location	West	East
Plant	600–800	550–750
Surface Coal Mine	225–350	150–275
Underground Coal Mine	—	1,000–1,500
Underground Shale Mine	400–600	—
Integrated Plant Average	1,000/1,200*	850/1,900*

*Underground mining.

people in the plant on the output and the raw fuel input. In Table 7.12 representative ranges of mine and plant personnel are shown, derived primarily from estimates by companies proposing to set up synthetic fuel plants.[3] These figures should only be regarded as illustrative. For coal mining, the principal difference between the East and the West is in the amount of coal assumed to be mined for a given plant output. We have taken 6×10^6 tons/yr as typical for the East and 8×10^6 tons/yr as typical for the West. The average plant populations shown for the West are for surface coal mining and underground oil shale mining. The averages shown for the East are for surface and underground coal mining.

In Ref. 3 sanitary and potable water usage in the mine and the plant was estimated on the basis of 30 to 35 gallons per man per shift, with 25 to 30 percent consumed. The recovered wastewater was assumed to be reused within the complex after treatment. The service and fire water usage for the mine was taken to be one and a half times the sanitary and potable water usage with no recovery, and for the plant it was taken to be two times the sanitary and potable water usage with about two thirds of the wastewater recovered. If these figures are appropriately weighted, there would be a net consumption of 25 gal/day per person employed in an integrated plant for sanitary, potable, service, and fire water needs. For 1,000 people this would amount to a consumption of 17 gal/min (8.5×10^3 lb/hr).

In any satellite town associated with an integrated synthetic fuel plant, the water use may be expected to be principally for domestic, commercial, and public consumption, with minimal requirements for the industrial portion of total consumption normally taken into account in municipal usage. Moreover, we may assume that all sewage water effluent in any such municipality will be treated for reuse within the mine-plant boundaries. Figures on per capita water use vary widely over the United States, with the range generally between about 100 and 175 gal/day per capita.[18] The consumed water may be between 30 and 40 percent of the usage. Generally, the higher usage and higher consumption are associated with the more arid areas, although this is not always the case when control and conservation are practiced. The total rate of water consumption for a satellite town will not be estimated here since this would involve projecting its population, which may not be entirely related to the mine-plant complex. However, this need must always be borne in mind in any overall planning.

Within the boundaries of the conversion plant, water will be needed for dust control at a number of points. These points are similar to those described for the mine and include transfer areas, active storage, and surge bins. Less water will be required for this purpose in the plant than in the mine. In Table 7.13 estimated ranges of water requirements are given for the control of dust from eastern and western coals in the nominal amounts used for the standard size plants. Also listed are the oil shale water requirements as scaled from the designs of Refs. 6 and 7. The lesser unit amounts for the Paraho designs may be a result of the rather large size shale lumps that are processed.

If the feed water to the complex comes from a dirty or turbid surface stream, then a settling basin would be used at the supply source to settle out the suspended matter prior to pumping the water to an on-site reservoir for plant and mine use. In the settling basin, if one is used, and in the on-site storage reservoir, water will be lost through evaporation. This loss is a penalty in water consumption chargeable against the integrated plant. Of course, in any water holding area a credit for rainfall should also be noted. Of interest, therefore, is the difference between the evaporation rate and the rainfall. We have already observed that in the Appalachian and Illinois basins the mean annual precipitation generally ex-

Table 7.13
Water consumed in dust control in synthetic fuel plants.

Process	Coal/Shale Processed tons/day	Water Consumed gal/min	lb water/10^3 lb fuel
Eastern Coal Conversion	18,000	24–36	8–12
Western Coal Conversion	24,000	32–48	8–12
TOSCO II	73,000	122	10
Paraho Direct	87,000	58	4
Paraho Indirect	100,000	67	4

ceeds the annual evaporation rate. Only in the western coal mining regions, therefore, are estimates of evaporative losses from water holding areas needed.

In Ref. 3, it is assumed that peak usage and surge periods can be accounted for by designing any settling basin for a throughput 50 percent greater than the mean annual consumption. Although a settling basin may not be used, an on-site reservoir will almost always be needed. In the same reference, the capacity of the reservoir is assumed to be sufficient to hold a one week supply of water for the mine and plant. The basin depth is selected to be 3 ft with a holding time of one day, while the reservoir depth is arbitrarily chosen to be 21 ft for the holding time of seven days. The storage time per unit water height is then the same for both holding areas and with the same throughput the evaporative losses would be the same at a given location. Table 7.14 shows the annual pond evaporation rates minus the precipitation rates in the western coal mining regions. Also shown in the table is the fraction of the total water consumption in the mine-plant complex that would be evaporated in either the reservoir or the settling basin for the design conditions given above: the fraction of plant water evaporated is directly related to the evaporation-minus-precipitation rate. The fractional water consumption may therefore be scaled to any location, knowing the net evaporation rate for that location.

From the results of Table 7.14, it can be seen that the percent water evaporated will range from a low of about 0.25 percent in North Dakota for a reservoir only, to a high of about one percent in New Mexico for both a settling basin and a reservoir. This would represent a water loss in the range of 5 to 50 gal/min for total integrated plant consumptions between 3×10^6 and 7×10^6 gal/day.

Table 7.14
Annual pond evaporation rates minus precipitation rates in western coal mining regions and percent of total integrated plant water consumption evaporated in an on-site reservoir or settling basin.

State	Pond Evaporation Minus Precipitation (in/yr)	Percent of Plant Water Evaporated in Reservoir or Basin
New Mexico	53	0.46
Wyoming	40	0.35
Montana	35	0.30
Colorado	33	0.29
North Dakota	30	0.26

7.5 A Summary of Water Stream Quantities

A number of mining, disposal, and other water requirements have been estimated in this chapter. Most are not large except those for flue gas desulfurization, ash disposal, and spent shale disposal. The reader is cautioned to carefully interpret the summarized values given, since almost all the quantities are for "typical" examples and not general cases. On the other hand, an important point to be made in terms of quality is that most of the large mining and disposal water needs can be met with low quality water.

Table 7.15 summarizes the mining and disposal water requirements for the standard size coal conversion plants considered, assuming a feed of either 24,000 tons/day of low sulfur western coal or 18,000 tons/day of high sulfur eastern coal. Most of the differences in requirements between the East and the West are in direct proportion to the difference between the two assumed coal feed rates.

Bottom ash disposal is seen to be a large consumer of water. However, the amount of water consumed depends directly upon the ash content of the coal, and in our illustrative example we have arbitrarily selected both coals to have a 10 percent ash content. The quantity of water listed for fly ash disposal and flue gas cleaning is mostly for flue gas desulfurization. The values shown do not repre-

Table 7.15
Mining and disposal water requirements in gal/min for integrated coal conversion plants converting 24,000 tons/day of 10 percent ash, 0.9 percent sulfur western coal or 18,000 tons/day of 10 percent ash, 3 percent sulfur eastern coal.

Purpose	West	East
Surface mining	35–110	—
Underground mining	—	100–300
Coal preparation	45–60	30–45
Bottom ash disposal	320–640	240–480
Subtotal	400–810	370–825
Fly ash disposal and flue gas cleaning	350*/660†	490*/475†
Land reclamation**	120	—

*Estimate for 15 percent of coal burned in boilers; bottom ash requirement is 90 percent of range in table.
†Estimate for Synthane process where char is burned in boilers; bottom ash requirement is 20 percent of range in table.
**New Mexico only.

Table 7.16
Mining and disposal water requirements in gal/min for integrated oil shale plants producing 50,000 barrels/day of synthetic crude.

Purpose	Paraho Direct*	Paraho Indirect*	TOSCO II†
Underground mining	350	400	390
Shale preparation	150	170	170
Shale disposal and revegetation	890	2,320	2,780
Total	1,390	2,890	3,340

*30 gal/ton shale.
†35 gal/ton shale.

sent a range. The first figure is an estimate for a plant in which about 15 percent of the feed coal is burned in the boilers to generate steam. The second figure is an estimate for a Synthane plant in which 4,000 tons/day of gasifier char from a western coal or 3,000 tons/day from an eastern coal are burned to generate steam.

The difference between the water requirements for the low and the high sulfur coals was obtained by using the unit requirements given in Table 7.9. Simple ranges cannot be given for the flue gas scrubbing requirement, so the values shown must be interpreted as typical cases. Moreover, much of the water consumed in scrubbing goes to saturating the flue gas and small changes in the flue gas temperature can lead to large differences in water consumed. Finally, the scrubbing water is not included in total consumption since, like land reclamation, it is not a requirement for every plant.

Table 7.16 summarizes the corresponding mining and disposal requirements for oil shale plants. Here there are marked differences between different designs, but again, care should be exercised in drawing general conclusions. The main reason for the differences lies in the assumed method of spent shale disposal. In the TOSCO II design the spent shale is both compacted and cemented with water.

Table 7.17
Service and other water requirements in gal/min for integrated coal conversion plants converting 24,000 tons/day of western coal and 18,000 tons/day of eastern coal.

Purpose	West	East	
	Surface	Surface	Underground
Sanitary, potable, service usage	15–20	10–20	25–40
Plant dust control	30–50	25–35	25–35
Evaporation	5–50	—	—
Total	50–120	35–55	50–75

Table 7.18
Service and other water requirements in gal/min for integrated oil shale plants
producing 50,000 barrels/day of synthetic crude.

Purpose	Paraho Direct	Paraho Indirect	TOSCO II
Sanitary, potable, service usage	20	25	20
Plant dust control	60	65	120
Evaporation	20	35	35
Total	100	125	175

In the Paraho designs the spent shale is simply compacted dry. The water consumption shown for the Paraho designs is mainly for revegetation, whereas in the TOSCO design it is largely for cementation. Whether the lower water consuming scheme is environmentally acceptable must still be determined.

The water requirements for service and other uses including evaporative losses are relatively small, as may be seen in Tables 7.17 and 7.18. When taken together, however, the total requirement is large enough not to be neglected in any integrated plant water balance.

Chapter 7. References

1. Nunenkamp, D. C., "Coal Preparation Environmental Engineering Manual," Report No. EPA-600/2-76-138, Environmental Protection Agency, Research Triangle Park, N.C., May 1976.

2. Fluor Utah, Inc., "Economic System Analysis of Coal Preconversion Technology, Volume 2 Characterization of Coal Deposits for Large-Scale Surface Mining," Report No. FE-1520-2, Energy Res. & Develop. Admin., Washington, D.C., July 1975.

3. Gold, H., Goldstein, D. J., Probstein, R. F., Shen, J. S., and Yung, D., "Water Requirements for Steam-Electric Power Generation and Synthetic Fuel Plants in the Western United States," Report No. EPA-600/7-77-037, Environmental Protection Agency, Office of Energy, Minerals & Industry, Washington, D.C., May 1977.

4. Grim, E. C. and Hill, R. D., "Environmental Protection in Surface Mining of Coal," Report No. EPA-670/2-74-093, Environmental Protection Agency, Cincinnati, Ohio, Oct. 1974.

5. "Ventilation and Dust Control Papers," *Second Symp. on Underground Mining,* pp. 325–375, National Coal Association, Washington, D.C., 1976.

6. Colony Development Operation, "An Environmental Impact Analysis for a Shale Oil Complex at Parachute Creek, Colorado, Part 1—Plant Complex and Service Corridor" (also, corrected water system flow diagram, personal communication), Atlantic Richfield Co., Denver, Colorado, 1974.

7. McKee, J. M. and Kunchal, S. K., "Energy and Water Requirements for an Oil Shale Plant Based on Paraho Processes," *Quarterly Colorado School of Mines* **71** (4), 49–64, Oct. 1976.

8. Western Gasification Co., "Amended Application for Certificate of Public Convenience and Necessity," Federal Power Commission Docket No. CP73-211, 1973.

9. Chu, T-Y., Nicholas, W. R., and Ruane, R. V., "Complete Reuse of Ash Pond Effluent in Fossil-Fueled Power Plants," Fiche No. 19, Paper No. 17f, AIChE 68th Annual Meeting, Los Angeles, Calif., Nov. 1975.

10. DiGioia, A. M. and Nuzzo, W. L., "Fly Ash as a Structural Fill," ASCE Preprint No. JPG-70-9, ASME-IEEE Joint Power Generation Conference (ASCE Participating Society), Pittsburgh, Penn., Sept. 1970.

11. Jones, J. W., "Disposal of Flue-Gas-Cleaning Wastes," *Chemical Engineering* **48** (4), 79–85, Feb. 1977.

12. National Academy of Sciences, *Rehabilitation Potential of Western Coal Lands.* Ballinger Publishing, Cambridge, Mass., 1974.

13. Aldon, F. E., "Techniques for Establishing Native Plants on Coal Mine Spoils in New Mexico," *Proc. Third Symposium on Surface Mining and Reclamation,* Vol. I, pp. 21–28, National Coal Association, Washington, D.C., 1975.

14. Schmidt-Collerus, J. and Bonomo, F. S. (eds.), *Proc. First Symp. on Management of Residuals from Synthetic Fuels Production,* Univ. of Denver, Denver Research Institute, Denver, Colorado, 1976.

15. Development Engineering, Inc., "Field Compaction Tests, Research and Development Program on the Disposal of Retorted Oil Shale Paraho Oil Shale Project," Report No. OFR78-76, Bureau of Mines, Dept. of the Interior, Washington, D.C., Feb. 1976.

16. Metcalf & Eddy Engineers, "Water Pollution Potential from Surface Disposal of Processed Oil Shale from the TOSCO II Process," Vol. I, Report to Colony Development Operation, Atlantic Richfield Co., Grand Valley, Colorado, Oct. 1975.

17. U.S. Department of the Interior, "Final Environmental Statement for the Prototype Oil Shale Leasing Program," Vol. I, U.S. Gov't. Printing Office, Washington, D.C., 1973.

18. Metcalf & Eddy, Inc., *Wastewater Engineering.* McGraw-Hill, New York, 1972, pp. 25–26.

8
Water Treatment

8.1 Water Treatment Plants

In the preceding chapters the various quantities of water required to operate synthetic fuel complexes and the qualities of the effluent process streams have been characterized. Influent water quality requirements, however, have not yet been specifically defined. Implicit in all these estimates of net water needs is that no water streams leave the mine-plant boundaries. All effluent streams are recycled or reused within the mine or plant after any necessary treatment. Since all of the conversion plants are net consumers of water, total reuse is of course possible. Water leaves the plant as vapor, as hydrogen in the hydrocarbon products, and as occluded water in the solid residues. The water treatment plants are not designed to return flows to receiving waters. Returning water to a source is not economic when the water must be cleaned to a quality equal to or better than the source water, to meet environmental constraints.

There are three main sources of water supply for synthetic fuel plants: surface water, groundwater, and municipal treatment plant effluent. The surface waters—rivers, streams, or lakes—are fresh and of good quality. The groundwater sources may be either fresh or brackish. If the brackish water is drawn from deep consolidated aquifers it will also be highly mineralized. Municipal effluent would come from any town allied with the mine-plant complex, although by itself it might not be sufficient to meet the needs of the entire plant.

Even when the water comes from the same type of source, no two waters will have the same quality. For example, the quality of water from rivers and streams may vary significantly during the year. On the other hand, the chemical composition of water from a given well at a particular location generally shows little variation with time, but the quality is spatially variable even in the same general location. The extent of the pretreatment facilities needed for the plant source water will depend on the chemical composition as measured by such water quality parameters as hardness, alkalinity, silica content, salt content, total dissolved solids, pH, and BOD.

A main difference between groundwaters is whether the water is brackish or fresh. Brackish water has a salt content greater than about 1,500 mg/l but less than about 10,000 mg/l. Large quantities of brackish groundwater are found in the arid western coal regions where a limited fresh water supply could lead to the selective use of brackish water for synthetic fuel plants. If brackish water is to be used it must first be desalted, and this is an added source water cost.

Among the most suitable means for brackish water desalting are the membrane processes of reverse osmosis and electrodialysis.[1-4] In electrodialysis, salt is removed through membranes by a direct current electric field. Reverse osmosis is much like filtering; hydrostatic pressure forces fresh water to pass through a membrane, leaving the dissolved material behind in a concentrated solution. Although the technologies are commercially available, they are not as yet wholly standard for water supply treatment. Moreover, they tend to be expensive, with representative costs of $1/10^3$ gal or more for brackish water desalting. Because brackish water is not seen as a primary water source for synthetic fuel plants its use will not be considered here, though it should always be kept in mind as an alternative or conjunctive supply. For the same reason we shall not detail the use of municipal effluents as a supply, though this water could also be expected to serve as a supplemental source in some instances.

In the discussion which follows, it is assumed that the water source is a relatively clean surface water. Even under this condition the water management scheme will differ among locations, depending upon differences in quality as well as available quantity. In any synthetic fuel plant, there are three main uses of water, each requiring a particular quality: high quality for the process, medium quality for cooling, and low quality for disposal and mine uses. Figure 8-1 represents one simplified water reuse scheme. The scheme assumes that the effluent from the process is of low quality and insufficient to meet all the plant's

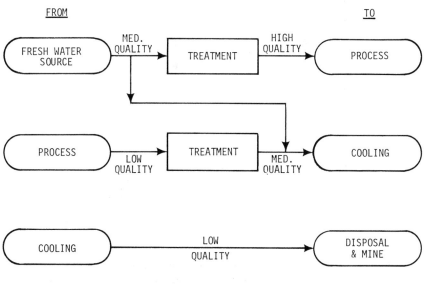

Figure 8-1
Simplified water reuse scheme.

cooling needs. For all the liquefaction processes and for the low and intermediate
temperature coal gasifiers, the process condensate is dirty. However, high temperature gasifiers such as Koppers-Totzek and Bi-Gas and the CO_2 Acceptor
process produce quite clean condensates. The scheme further assumes that the
supply is a fresh water, medium quality source. If the source of supply were of
poor quality and expensive, it might in fact be economical to take the medium
quality water resulting from treating the dirty process stream and feed it back for

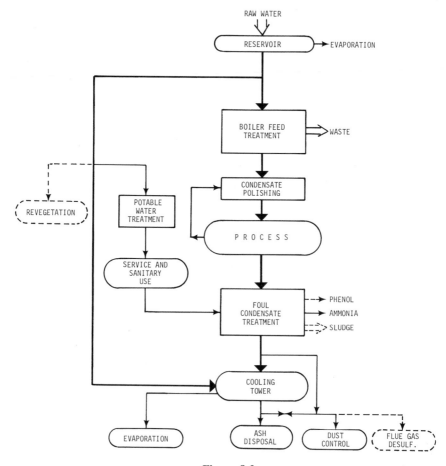

Figure 8-2
Water treatment flow diagram for coal conversion plant generating dirty process water.
(Dashed boxes for flue gas desulfurization and revegetation indicate the requirements are
not necessary for every plant.)

treatment to high quality boiler feed water. In addition to assuming a cheap, adequate, and relatively clean water supply, the discussion in this chapter will consider only coal conversion plants that produce a foul process condensate. The same treatment schemes would apply to shale conversion plants, but the water quantity distributions within the plant would be different.

Figure 8-2 is an amplification of Fig. 8-1 and represents a general water treatment scheme for a coal conversion plant generating dirty process water. The scheme is not unique but does contain the main components of any water treatment plant: boiler feed water preparation, process water cleanup, and cooling water. The three main streams are shown with heavy lines. Illustrated in Fig. 8-2 is the treatment of the raw water to boiler feed quality, the treatment of foul process condensate into makeup for the cooling tower plus water for dust control and, if needed, makeup for flue gas desulfurization. The use of raw water in the cooling tower is also shown. The dashed lines for the products from the foul condensate treatment indicate alternative schemes. In one the phenol may be recovered by solvent extraction followed by ammonia distillation. Alternatively, the ammonia separation may precede biological oxidation in which the organic matter is oxidized and produces a waste sludge. All three approaches may also be used. The treated foul condensate may in part be used for ash disposal or the cooling tower blowdown may by itself be used to meet this requirement.

In the following sections the treatment schemes required for boiler feed, dirty process condensate, and cooling water are described.

8.2 Boiler Feed Water

In all synthetic fuel plants large quantities of high pressure steam are needed for the conversion processes. Correspondingly large amounts of water must be fed to boilers to generate this steam. For the 250×10^6 scf/day pipeline gas plants examined, the high pressure process steam requirements range between about $1,000 \times 10^3$ and $1,500 \times 10^3$ lb/hr or about 3×10^6 to 4.5×10^6 gal/day. The requirements for the 50,000 barrel/day coal liquefaction plants have a similar range, except for the Sasol synthesis plant, which uses about $2,500 \times 10^3$ lb/hr.

Water fed to boilers to generate high pressure steam must be of a very high purity. The reason for this is that a boiler will not function well if its tubes become coated with scale or corrode. To prevent scaling or corrosion, as much dissolved material as possible must be removed from the feed water. Even the cleanest of natural streams must be treated to achieve the level of purity required. Treatment of such large quantities of water as a synthetic fuel plant might need could typically account for 25 percent of all the plant water treatment costs, the actual fraction depending on the quality of the source water. A high quality lake water will cost considerably less to clean up than a brackish groundwater.

The major part of boiler feed water treatment is the removal from solution of silica and calcium, magnesium, bicarbonate, and sulfate ions. These dissolved materials are the principal ones that lead to scale formation on the boiler tubes. In some high pressure boilers all the water fed to the boiler is vaporized and any non-volatile contaminants in the feed water that are not carried away as particles in the steam remain on the boiler tubes.[5] The scale on the tubes reduces the amount of heat transferred. The tubes may need cleaning as often as once a year but more usually as seldom as every five years. In other words, very pure water is needed to keep a high pressure boiler running smoothly. The purification of makeup water taken from a surface source is most often accomplished either by lime or lime-soda softening followed by ion exchange deionization, or by ion exchange alone. Not only is the makeup water deionized as completely as possible, but all the condensate returned to the boiler undergoes ion exchange treatment to ''polish'' it.

In lower pressure boilers, about 98 percent of the feed water is vaporized. The non-vaporized water, which is more concentrated in dissolved material than the feed, is blown down; that is, it is released as liquid from the boiler. The quality of the blown down water is usually not worse than the raw water entering the plant, so it is mixed with the raw water and treated for makeup. A boiler with blow-down can accept less pure water than that needed for a once-through boiler. Most of the boilers in synthetic fuel plants will have a blowdown.

Apart from silica, the principal dissolved materials to be removed from boiler feed water are in ionic form in water. Suggested limits for the contaminant concentrations in boiler feed water have been developed. The simple weight unit mg/l is used as a measure of total dissolved solids and silica concentration but is not convenient for the measurement of ion or ionized radical concentrations. One reason is that in mg/l the sum of the cations (positive radicals) will not equal the sum of the anions (negative radicals), so that the analysis cannot readily be checked to see that the water is electrically neutral as it must be. Furthermore, a simple weight expression of concentration does not allow for ready calculation of quantities of chemical precipitants or other chemicals. Ionic concentrations are therefore usually expressed in terms of their combining, or equivalent, weights. Two expressions are used: milliequivalents per liter, written ''meq/l,'' or milligram equivalents of calcium carbonate per liter of water, written ''mg/l as $CaCO_3$.'' The equivalent weight of an ion is equal to its atomic weight divided by

Table 8.1
Equivalent weights of common ions and ionized radicals found in water.

Cations			Anions		
Name	Symbol	Equivalent Weight	Name	Symbol	Equivalent Weight
Ammonium	NH_4^+	18.0	Bicarbonate	HCO_3^-	61.0
Calcium	Ca^{2+}	20.0	Carbonate	CO_3^{2-}	30.0
Cuprous	Cu^+	63.5	Chloride	Cl^-	35.5
Cupric	Cu^{2+}	31.8	Fluoride	F^-	19.0
Ferrous	Fe^{2+}	27.9	Hydroxide	OH^-	17.0
Ferric	Fe^{3+}	18.6	Iodide	I^-	126.9
Magnesium	Mg^{2+}	12.2	Nitrate	NO_3^-	62.0
Potassium	K^+	39.1	Sulfate	SO_4^{2-}	48.0
Sodium	Na^+	23.0			

its valence and for an ionized radical its electrical charge divided by its molecular weight. For convenience we have shown in Table 8.1 the equivalent weights of the most common ions and ionized radicals found in water, both of which we shall refer to as ions. The formulas for converting between the different units are

$$mg/l = \text{equivalent weight} \times meq/l, \qquad (8.1)$$

$$mg/l \text{ as } CaCO_3 = 50 \times meq/l. \qquad (8.2)$$

Alkalinity of water is a measure of its ability to absorb strong mineral acid with a limited decrease in pH, in other words, a measure of its capacity to neutralize acids. Although many ions contribute to alkalinity, the three ions whose concentration is determined as alkalinity are the hydroxyl, carbonate, and bicarbonate ions. Alkalinity as a measure of ion concentration is also conventionally expressed in terms of mg/l as $CaCO_3$. Hardness is a measure of the concentration of calcium and magnesium, and hence the removal of these ions from water is termed "softening." Some suggested tolerances in boiler feed water for alkalinity, hardness, total dissolved solids, and silica are given in Table 8.2, as modified from Ref. 6. The limits are seen to be quite low, particularly for the higher pressure range.

To prevent corrosion, dissolved oxygen must also be reduced to very low levels. A maximum concentration of 0.007 mg/l has been recommended. Suspended and dissolved copper and iron should also be reduced down to 0.01 mg/l

Table 8.2
Suggested tolerances for boiler feed water.

Boiler Pressure (psig)	Total Dissolved Solids* (mg/l)	Alkalinity* (mg/l as $CaCO_3$)	Total Hardness (mg/l as $CaCO_3$)	Silica* (mg/l SiO_2)
300–400	60	12	0.3	1.8
1,000–1,500	20	0	0.0	0.04

*For 2 percent blowdown.

for the higher pressure range, and 0.05 and 0.025 mg/l, respectively, for the lower pressure range.[6]

Ion exchange sometimes preceded by lime or lime-soda softening is the principal method for obtaining high purity boiler feed water from a raw water source. In lime treatment, lime (calcium hydroxide) is added to the water to cause the precipitation of magnesium hydroxide and calcium carbonate.[7-9] Both precipitation reactions require adjustment of the pH value into the alkaline range, done by adding excess lime. After precipitation the water contains hydroxyl ions and calcium from the excess lime. Soda ash (sodium carbonate) may be added in a second stage to reduce the calcium content by precipitating out calcium carbonate.

In treating cold water, the chemical reactions are too slow to be carried to completion. The practical limits of cold lime precipitation softening are usually 30 to 40 mg/l of calcium carbonate and 10 mg/l of magnesium as $CaCO_3$.[8] If the water is to be used hot, as in a boiler, then it can be heated to over 200°F before the lime is added.[9-11] This speeds up the reactions and increases the removal rate.

Silica is precipitated as a complex with magnesium hydroxide. If enough magnesium is not present to precipitate the desired amount of silica, more magnesium may be added. Removal of silica is more complete when the water is heated before treatment.

The precipitated material settles in a clarifier, which is basically a large tank in which the water resides long enough to permit the solid particles to settle. Often precipitant settling is aided by agitation to enhance flocculation of the material and by adding chemical agents to enhance coagulation of the suspended material. The underflow from the clarifier, if not sufficiently concentrated, may be sent to a thickener, which is similar to a clarifier except that its purpose is to provide a dense underflow. A thickener would be deep enough to hold a greater volume of sludge with enough weight to compress the bottom of the sludge blanket and squeeze water from it. Even from a thickener the sludge is still usually removed by some means of liquid handling. The sludge may require further dewatering by filtration or centrifugation. The clarified water from the overflow will be filtered prior to ion exchange treatment. Generally, the water will first be filtered by gravity filtration through a sand or a mixed media filter, followed by a polishing filter. Polishing filters pass a high flow rate of water and pressure filters are usually used. (Further details on the procedures for suspended solids removal may be found in any text or review on wastewater treatment, for example, Refs. 7 to 9.)

The source water, whether lime treated or not, will be filtered before passing through ion exchange units for deionization and silica removal. Ion exchange is a process in which the unwanted ions in solution are exchanged with innocuous ions within an insoluble ion exchange material. (Detailed descriptions of ion exchange appropriate to our discussion may be found, for example, in Refs. 7 and 9 to 11.) The most important class of solid ion exchangers in use today for water treatment are synthetic organic resins, produced typically as millimeter size spheres. These resins contain charges fixed within their solid structure and mobile charges that can be exchanged. One class of resins contains exchangeable positive ions and another class exchangeable negative ions. The supply of exchangeable ions within the resin is, however, limited and after the resin is exhausted of these ions it must be regenerated for reuse. The resins we shall consider remove the unwanted ionic species by recharging them with hydrogen (H^+), sodium (Na^+), and hydroxyl (OH^-) ions. After exhaustion the resins are regenerated, respectively, with a concentrated solution of sulfuric acid (H_2SO_4), sodium chloride (NaCl), and sodium hydroxide (NaOH). The ion in the regenerant solution that is not used to regenerate the resin, for example, the Cl^- ion in an NaCl solution, combines with the unwanted ion removed from the resin, for example, a Ca^{++} ion, to give a concentrated waste solution of $CaCl_2$. The resins that have been developed are quite resistant to physical and chemical degradation and may be used and reused over long periods of time. For water softening less than one percent of the resin needs to be replaced each year.

Ion exchange treatment of boiler feed water is most frequently carried out batchwise, with resin loaded into a cylindrical vessel. In a typical cycle the water to be treated is uniformly introduced at the top over the vessel cross-section. It then flows down through the resin bed and is removed through a collecting system beneath the bed. After the bed is exhausted, as evidenced by the contaminant ions appearing in the effluent, the feed water is shut off. The bed is then usually backwashed to remove accumulated suspended matter, following which it is regenerated. The regeneration may be carried out either in the direction the water flows through the bed, which is current practice in the United States, or counter to it. Countercurrent regeneration is more efficient in the utilization of the regenerant; if any unwanted ions are left in the resin, they are left at the top of the bed where the feed enters during the exchange cycle and not at the bottom of the bed where they can leak out with the treated water at the beginning of the exchange cycle. The principal problem in countercurrent regeneration is to prevent bed fluidization. Following regeneration, the excess regenerant is washed off from the bed before using it again. The volume of liquid waste may be estimated to be 5 to 15 percent of the volume flow through the ion exchange system, with the lower figure for countercurrent regeneration.

Ion exchange is a standard operation that has been used extensively for boiler feed water treatment and water treatment in general for many years. But there can be no standard ion exchange treatment procedure applicable to the boilers in every synthetic fuel plant: the quantities of boiler feed water to be treated to high purity in a synthetic fuel plant are extremely large, and moreover, no two source waters have the same contaminants or the same levels of contamination. Innovative utilization of ion exchange resins must be employed to minimize the amount of regenerant waste disposed and at the same time minimize the operating cost, almost all of which is in the regenerant chemicals.

Several resins have been found particularly useful in the design of ion exchange treatment units for boiler feed water. A "weak acid" resin is used to exchange positive ions in the water with H^+. The resin removes only that part of the total positive ions equivalent in amount to the bicarbonate alkalinity. This resin is particularly good for removal of Ca^{++} and Mg^{++} and less satisfactory for removing Na^+. When H^+ has been substituted for Ca^{++} and Mg^{++}, carbonic acid (H_2CO_3) is formed. This decomposes to free carbon dioxide, which then is released in a degasifier. A great advantage of weak acid resins is the ease of regeneration with sulfuric acid. Little more than the theoretically required amount of sulfuric acid is needed to displace the unwanted ions in the resin.

A "strong acid" resin is used to replace all of the positive ions with H^+. Regeneration of strong acid resins with H_2SO_4 may be very inefficient. With countercurrent regeneration, about 1.5 to 2 times the theoretical amount of H_2SO_4 is required for regeneration. In those designs employing weak and strong acid resins, the regenerant acid is poured first through a strong acid resin and then through a weak acid resin to enable a high use of the acid.

A "softening" resin is used to replace Ca^{++} and Mg^{++} with Na^+. It is regenerated with NaCl, but the regeneration is not very efficient. About twice the theoretically required amount of salt is needed to displace the Ca^{++} and Mg^{++}, even with countercurrent regeneration.

A "weak base" resin is used to replace $SO_4^=$ with OH^-. It cannot remove weakly dissociated carbonic acid from the alkalinity or the very weak silicic acid (H_2SiO_3) formed from the silica that passes through a positive ion exchange unit. A weak base resin is easily regenerated with NaOH. This resin is used following a strong acid resin to produce a demineralized water.

A "strong base" resin is used to replace both the weakly dissociated and the strongly dissociated acids. It may also be used to remove silica by replacing $HSiO_3^-$ with OH^-. It is regenerated with NaOH. The regeneration is inefficient; thus to obtain a high use of the chemical, the NaOH is passed from the strong base to the weak base resin.

A "mixed bed" is a mixture of strong acid and base resins designed to polish demineralized water without large changes in pH. This treatment is necessary for high pressure boilers.

To illustrate the differences in selecting resins when treatment costs are to be minimized, consider the treatment to boiler feed water quality of a source water that is a quite clean, slightly hard lake water. The analysis of the water, taken from a lake in North Dakota,[12] is given in Table 8.3. The source water is assumed to be treated to meet the requirements in Table 8.2 for both ranges of boiler pressure shown there. Examples of gasification processes where steam in the lower pressure range (less than about 400 psig) would be used are Lurgi,

Table 8.3
Analysis of a clean, slightly hard lake water.

	mg/l	mg/l as $CaCO_3$	meq/l
Ca^{2+}	49	123	2.45
Mg^{2+}	19	78	1.56
Na^+	59	129	2.57
HCO_3^-	180	148	2.95
SO_4^{2-}	170	177	3.54
Cl^-	9	13	0.25
Silica	7	—	—
Suspended Solids	2	—	—
Dissolved Solids	428	—	—

Koppers-Totzek, and CO_2 Acceptor. Steam in the higher pressure range (about 1,000 psig) would be needed for Synthane and Hygas plants. For the gasifiers in the liquefaction plants, steam in the lower pressure range would generally be used. Typically 80 to 90 percent, or a range of 800×10^3 to $1,300 \times 10^3$ lb/hr (1,600 to 2,600 gal/min), of the high pressure process steam requirement would be produced from treated water drawn from the source. The other 10 to 20 percent of the steam would be produced from methanation water or other clean water returned from the process, requiring only "polishing" treatment.

To meet the boiler feed water requirements in the lower pressure range, the relatively standard approach of lime-soda softening followed by strong acid, weak base demineralization (Fig. 8-3) may be used to treat the quality of lake water given in Table 8.3. The effluent from the lime-soda treatment will be reduced to 12 mg/l of Ca^{++} and 2.5 mg/l of Mg^{++}, corresponding to 30 and 10 mg/l, respectively, as calcium carbonate. Most of the bicarbonate will be removed and the silica reduced to around 1 mg/l. The other dissolved ions will pass through essentially unchanged. The removal of the remaining dissolved positive ions takes place in the strong acid ion exchange unit, while the weak base unit removes the negative ions. Sulfuric acid and sodium hydroxide are the regenerants. An illustrative cost[13] might be about $0.20/10^3$ gal for the lime-soda treatment and $0.50/10^3$ gal for the ion exchange demineralization. (The energy requirements for the treatment are quite small.) The regenerant waste load would be in the range of 40×10^3 to 60×10^3 lb/hr (80 to 120 gal/min) depending upon the total steam requirement of the plant. Although some water is occluded in the lime treatment sludge, which is not easily recoverable, the amount is small and may be neglected.

We next consider the treatment of the same water, but this time to boiler feed

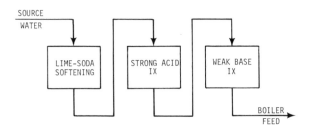

Figure 8-3
Lower pressure boiler feed water treatment of clean
lake water.

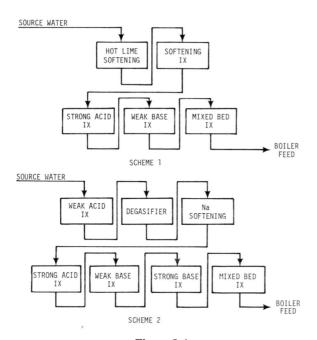

Figure 8-4
Higher pressure boiler feed water treatments of clean
lake water.

quality for the higher pressures. Two schemes shown in Fig. 8-4 are considered.[13] Scheme 1 is a more conventional one in which hot lime treatment precedes ion exchange. This scheme is closer to the procedure outlined above for the lower pressure boiler feed water treatment. The main differences are that a softening ion exchange follows the lime treatment in which Ca^{++} is replaced with Na^+, and a mixed bed polishing step is added at the end. The silica and hardness are low enough after the exchange on the softening resin, but the Na^+ has been increased. Demineralization is carried out in a strong acid exchange followed by a weak base exchange, similar to the scheme in Fig. 8-3.

Scheme 2 shown in Fig. 8-4 has more steps, but regenerant chemicals are more efficiently used. Here the water is first treated with a weak acid ion exchange resin to remove Ca^{++}, Mg^{++}, and some Na^+ and to replace these ions with H^+. From bicarbonate the hydrogen ions release CO_2, which is then removed in the degasifier. Additional removal of Ca^{++} is required on a softening resin. As before, demineralization requires a strong acid resin followed by a weak base resin. The strong base resin removes the silica. As with all ion exchange schemes for very high pressure boiler feed treatment, a mixed bed system is used for final polishing. Because it uses less regenerant chemicals, this scheme is cheaper to operate, though its capital cost is higher. A return condensate polishing system would be required, but this is inexpensive. An illustrative cost for this scheme might be about $0.90/$10^3$ gal, which is less than 30 percent more than that for the lower pressure water treatment scheme, despite the much higher purity water produced. As for the lower pressure scheme, the energy requirements are negligible and the volume of liquid regenerant wastes are about the same.

8.3 Process Condensate

For 250×10^6 scf/day pipeline gas plants, dirty process condensates range from a low of about 300×10^3 lb/hr for the Hygas process to about $1,100 \times 10^3$ lb/hr for the Lurgi process; the Synthoil dirty condensate is about 400×10^3 lb/hr and the SRC condensate 700×10^3 lb/hr. The Sasol synthesis plant, which uses Lurgi gasifiers to produce $1,000 \times 10^6$ scf/day of synthesis gas, naturally produces larger quantities of dirty process water. For our purposes the range of dirty condensate may be taken to be 300×10^3 to $1,100 \times 10^3$ lb/hr (600 to 2,200 gal/min) for the corresponding boiler feed rates considered in the last section.

All of the foul process condensates recovered from the liquefaction and low and intermediate temperature gasification plants have a high chemical oxygen demand (COD), up to 30,000 mg/l, and a high biochemical oxygen demand (BOD), up to 20,000 mg/l. They are heavily contaminated with dissolved organic matter, ammonia, and acid gases. From the analyses and discussions in Chapters 5 and 6, the SRC, Synthoil, and Synthane processes are expected to produce dirtier waters than, for example, the Hygas and Lurgi processes. However, they all share the need to remove large quantities of phenol, roughly in the range of 3,000 to 10,000 mg/l, and large quantities of ammonia in the range of 3,000 to 16,000 mg/l or higher. The exact amounts were shown to be very dependent on the process and the coal.

To remove or destroy phenol and other organic matter, many possible procedures exist including solvent extraction, adsorption on carbon or synthetic polymers, biological or chemical oxidation, and others. These procedures may be used alone or in combination. Ammonia separation and recovery may also be accomplished in a number of ways, although distillation, relying on the volatility of ammonia, is usually a main feature of any method used for the concentrations considered here. Choosing the method or combination of methods to be used is a detailed economic question, whose answer shall not be attempted here. Several procedures are in use, or are considered likely candidates for use, in the water treatment plants of synthetic fuel complexes. These include phenol and organic matter destruction by biological oxidation, recovery by solvent extraction, and ammonia separation by distillation.

Ammonia Separation

Being volatile, ammonia and the acid gases carbon dioxide and hydrogen sulfide are not difficult to separate by fractional distillation.[15] When the water containing the dissolved gases is boiled, the vapor is more concentrated in the gases than was the liquid. If the vapor is condensed and reboiled, the vapor is still more concentrated. This is the principle of fractional distillation.[16] However, separation of the gases from the water is not enough. The gases must be separated into ammonia and acid gases. Ammonia is a salable commodity and should be collected in a salable form. Hydrogen sulfide cannot be vented to the atmosphere but must be collected and sent to a sulfur recovery plant.

One process for separating out ammonia and the acid gases, proprietary to Lurgi Gesellschaft, is pictured in Fig. 8-5.[14] The vapor pressure of carbon

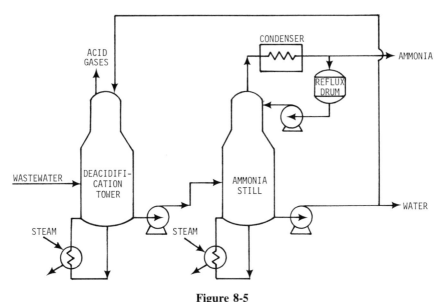

Figure 8-5
Distillation procedure to separate process wastewater into water, ammonia, and acid gas streams.

dioxide and hydrogen sulfide over the process condensate is more than the vapor pressure of ammonia. If clean water is poured down the distillation column, called refluxing, it is possible to operate so that carbon dioxide and hydrogen sulfide go overhead as a vapor with very little ammonia and water, while water, ammonia, and some small amount of carbon dioxide leave as a liquid at the bottom. Water is separated from ammonia in a second distillation column. Approximate calculations suggest that about 100 lb of clean water are circulated for every 100 lb throughput.[13] This puts a high energy load on the ammonia still, and the energy required for the overall process is in the range of 25 to 30 lb steam/100 lb water treated (about 2.0×10^6 to 2.5×10^6 Btu/10^3 gal). In distillation procedures the energy requirement and equipment size are for the most part independent of the concentration in the feed. At $2/$10^6$ Btu the energy charge alone is quite high. It may be expected that half as much again would be required to amortize the capital cost. However, any charge calculated in this manner is somewhat misleading since ammonia is a salable commodity; any cost for its removal should be balanced against any income recovered from its sale.

In another process proprietary to U.S. Steel, ammonia is absorbed into phosphoric acid.[17] There exists a range of ammonia-to-phosphoric acid mole ratios over which ammonia is bound tightly enough for good absorption but still loosely enough to be stripped back out. The absorber is usually placed directly above the water stripping still. In the absorber ammonia is removed from the gas at close to atmospheric pressure. The desorber is operated at elevated pressure, which helps to release ammonia. Desorption is forced by steam heat. Aqua ammonia vapor at 10 to 20 weight percent leaves the desorber, is condensed, and then fractionated in a standard still at elevated pressure. This complete process, including stripping the water, requires about 20 lb steam/100 lb water treated (1.7×10^6 Btu/10^3 gal).

Biological Treatment

When biological oxidation is used to destroy the organic matter, ammonia separation will precede it because high concentrations of ammonia are toxic to the bacteria that biologically degrade the organic matter. Aerobic biological decomposition of organic matter in wastewater to a settleable cellular sludge, carbon dioxide, and water requires the presence of aerobic bacteria and dissolved oxygen. For most wastewaters, the concentration of bacteria yielded by the decomposition of the organic matter in a once-through reactor tank, into which air or oxygen is mixed, is not high enough to obtain an adequate rate of reaction.

Usually the sludge is separated from the slurry leaving the oxidation tank and most is returned to the tank, maintaining a high bacterial concentration in the reactor at all times. The balance of the sludge is disposed. This procedure is called the activated sludge process.

Another way to obtain high concentrations of bacteria is to grow the bacteria on a solid surface and to pass the wastewater in a thin film over the surface. This is the principle of a trickle filter. For background information on these processes see, for example, Refs. 7, 8, and 18.

The activated sludge process is widely used, mostly in municipal sewage treatment plants. Figure 8-6 shows a scheme appropriate for the treatment of coal conversion wastewater. To keep any changes in flow and concentration to the bioreactor as small as possible, the feed comes from a storage pond or tank ("equalization basin"). The bioreactor is one or more stirred basins arranged in series. Phosphorus, which is an essential nutrient for the bacteria, has to be added. The concentration of organic waste in the reactor is close to the concentration in the water leaving the reactor. Phenol is toxic to bacteria above about 500 mg/l, but by carefully distributing the concentrated feed the toxicity of high concentrations of phenol can be avoided.

The biological agents in municipal sewage plants are simple soil bacteria. Coke oven wastewater, which is related to coal conversion wastewater, has been treated by biological oxidation for some years, and the bacterial sludge from such a plant is more efficient.[19] These bacteria can be acclimated to a phenolic waste and can as such grow quicker than unacclimated bacteria. Bacteria deliberately

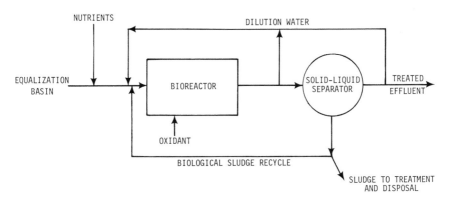

Figure 8-6
Part of an activated sludge biological oxidation process for the treatment of coal conversion wastewater.

mutated to acclimate them to phenol are also available and seem to be even more efficient.[20]

The oxidant can be air or oxygen. In municipal sewage plants air is the most common oxidant. To keep the sludge in suspension, the bioreactors must be stirred. If the stirrers are positioned to also pull air into a vortex in the tank, the oxidant addition is obtained for only a modest increase in energy consumption over that needed for stirring. However, municipal sewage may have only 200 mg BOD/l, compared to as much as 20,000 mg BOD/l in wastes from coal conversion plants. To use simple surface stirrers to pull air into the water will consume a great deal of energy, possibly in the range of 0.4 to 0.45 kwh/lb BOD removed. For 15,000 mg BOD/l removed this translates to 50 to 56 kwh/10^3 gal, almost equal to the energy requirement to evaporate water in a modern vapor compressor evaporator.

Oxygen, instead of air, is probably preferable for high strength wastes. In any case, most coal conversion processes make oxygen on site, so its use could be cheaper than the use of air. Furthermore, when oxygen is used the energy consumed is somewhat less, about 0.3 kwh/lb BOD removed. A disadvantage of oxygen, however, is that the waste stream heats up. When using oxygen the bioreactors are covered, thus preventing the large evaporation caused by nitrogen sweeping through the reactors. The wastewater temperature increases 1°F for roughly every 200 mg/l BOD removed. If 15,000 mg BOD/l are removed, this corresponds to a temperature increase of 75°F. Above a temperature of about 95°F the bacteria are inactive so that a rise in temperature of only about 15°F is permissible. The dilution stream shown in Fig. 8-6 has to be used and a cooling tower has to be put in the stream to force evaporative cooling. There is little experience with passing water of this type through a cooling tower. Bacteria and other growth will certainly occur when the water is oxygenated in the tower. Growth must not be prevented by bacteriocides, because the water does return to the bioreactor. Most probably the cooling tower would turn into a trickle filter, which would facilitate the process. Indeed, the cooling tower should perhaps be designed to combine evaporative cooling with bio-oxidation.[22] The dilution stream, which is circulated internally, is several multiples of the throughput. If the dilution stream can be a slurry taken before the clarifier, a good deal of the cost of building very large clarifiers could be saved. An approximate cost estimate for this system is about $0.02 to $0.025/lb BOD removed, or $2.5 to $3.1/$10^3$ gal for 15,000 mg BOD/l. The corresponding energy requirement is quite high, about 37 kwh/10^3 gal for 0.3 kwh/lb BOD removed.

Phenol Extraction

The extraction of phenol from the process condensate has the advantage of recovering the phenol at least for use as a fuel or for a product with an even higher value. Moreover, extraction may remove the necessity for biological treatment or at least minimize its cost and problems. If phenol is extracted, this will precede ammonia separation because phenol is acidic and tends to hold ammonia in solution preventing its evaporation. Extraction processes are usually complex, high in energy consumption, and expensive.

To extract phenol from the dirty process water, the water must be contacted with a solid that absorbs phenol on its surface or a liquid solvent that dissolves phenol preferentially over water. In the solvent extraction method considered here, the liquid that directly contacts the water must not mix with it. Solvents that have been used include benzene and kerosene. Part of the complexity of solvent extraction is the large number of solvents and mixtures of solvents that can be used, but the performance of many of these has still not been evaluated.[23]

One measure of the solvent quality is the "distribution ratio," k, defined as the ratio of the concentration of material extracted in the solvent to the concentration of the material in the water. For phenol extraction with benzene, $k = 2.3$, which is quite low. Two modern solvents are isopropyl ether ($k \sim 50$) and butyl acetate ($k \sim 70$). However, the distribution ratio for phenol is not the only important attribute of a solvent. In the case of dirty process condensates, the distribution ratio for the BOD must also be known in order to decide whether biological treatment is or is not required subsequent to extraction. This information is not available in the literature.

No solvent is completely insoluble in water, necessitating the removal of residual solvent after extraction. Something must also be done with the solution of phenol and other materials in the solvent. If the solvent is a common, cheap fuel like kerosene or light oil as produced by coal conversion, it may simply be burned with all of the extracted organic matter. A more usual procedure is to separate solvent and phenol by distillation. This means that two more essential attributes of the solvent are a high volatility relative to phenol and a low latent heat of vaporization.

Figure 8-7 shows an idealized solvent extraction process. The energy consumed, mostly to distill the solvent, is the product of the solvent flow rate and the solvent latent heat of vaporization. To show how the flow rate depends on the extractive ability of the solvent and the design, an ideal countercurrent extraction

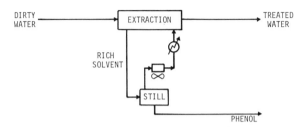

Figure 8-7
Idealized solvent extraction.

in n stages is assumed, in which the solvent and water are completely immiscible. From material balances, the relation between the number of stages, n, the desired extraction ratio, R, the distribution ratio, k, and the ratio of solvent flow rate, s, to water flow rate, w, for $sk/w > 2$ can be shown to be given approximately by

$$n \approx \frac{\ln R}{\ln (sk/w)}. \tag{8.1}$$

The extraction ratio R is the ratio of the concentration of the phenol in the dirty feed water to the concentration in the product water. The relation between R and the fractional removal, f, is given by

$$R = \frac{1}{1-f}. \tag{8.2}$$

For a 99.9, 99, and 95 percent phenol removal, R is 1000, 100, and 20, respectively.

Equation (8.1) shows that all else being equal, the number of stages decreases as the solvent rate is increased. With fewer stages the extraction column can be smaller and cheaper. However, the decrease in number of stages with increasing solvent rate is only logarithmic and is therefore slow. Moreover, since all of the solvent has to be evaporated to separate it from the phenol, increasing the solvent rate increases the cost of the still and the energy.

Every detailed engineering design requires an economic optimization that must be repeated for each solvent and for each fractional removal. However, a simple approximate calculation will be given. The cost of solvent extraction may be expressed by a simple cost equation similar to what was done for dry and wet coolers (for example, see Eq. 4.6). The cost of treating the water C (say, in $/lb water) is taken to be the sum of the amortized capital cost plus the energy cost.

The capital cost is assumed to be equal to the number of stages multiplied by an amortized capital cost coefficient k_c (also, say, in $/lb water). The energy cost is the product of the energy needed to distill the solvent, sL (L is the latent heat of vaporization), and the cost of heat energy k_e (for example, in $/Btu) divided by the water flow rate. The cost equation is then written

$$C = k_c n + k_e \cdot \frac{sL}{w} . \tag{8.3}$$

The number of stages may be expressed in terms of the solvent flow rate by Eq. (8.1). If this is done, and if the total cost C is optimized by differentiating Eq. (8.3) with respect to s/w and setting the result equal to zero, the optimum value of s/w is given by

$$\left(\frac{s}{w}\right)_{opt} \ln \left[k \left(\frac{s}{w}\right)_{opt} \right] = \frac{k_c \ln R}{2 k_e L} . \tag{8.4}$$

Equation (8.4) is a simplified and convenient equation for determining the optimum value of s/w for a chosen solvent (k and L chosen), a chosen extraction ratio, R, and given economic factors, k_c and k_e. More precise optimization will usually be required before the design is fixed, but if the detailed calculations are made around the value of s/w determined from Eq. (8.4), much time and effort may be saved. To illustrate the utility of Eq. (8.4), let us choose an amortized capital cost coefficient k_c of $1/10^3$ gal, an energy cost k_e of $2/10^6$ Btu, and a 99 percent phenol removal. With k varying from 2.5 to 75 and a latent heat of 170 Btu/lb, the optimum solvent to water ratio $(s/w)_{opt}$ varies from 1.25 to 0.21. In a preliminary design study[13] a value of $(s/w)_{opt} \approx 0.6$ was found for the same extraction ratio and latent heat as above with $k = 2.3$.

To achieve good contact and good separation between the water and solvent, there is probably a practical lower limit to the solvent-to-water ratio. Contacting means mixing followed by settling in a number of stages, while the solvent and water flow countercurrently. Thus a density appreciably different from that of water is another necessary attribute of the solvent. A reasonable example for a minimum value is $(s/w)_{min} = 0.1$. For a latent heat of vaporization of 170 Btu/lb, the minimum solvent-to-water ratio corresponds to an energy consumption of 140 × 10^3 Btu/10^3 gal of water. This may be compared with 850 × 10^3 Btu/10^3 gal for the optimum ratio of 0.6.

The two energy quantities corresponding to the minimum and optimum ratios are shown in Fig. 8-8 along with the energy consumed in biological oxidation. The line for biological oxidation is based on 0.4 kwh/lb BOD removed (4,000

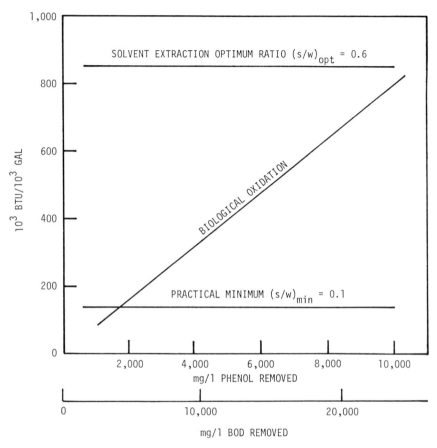

Figure 8-8
Energy for phenol removal by solvent extraction and biological oxidation.

Btu/lb BOD removed, with 1 lb phenol having 2.38 lb BOD). From the point of view of energy consumption, solvent extraction, which is independent of the concentration of contaminants, is most attractive at higher concentrations. With the best solvents, solvent extraction may become attractive for as little as 2,000 mg/l phenol, a level likely to occur in many coal conversion wastewaters.

Even if there is a practical lower limit to the solvent-to-water ratio, there is an advantage to very high distribution coefficients. With values of k from 50 to 100, sk/w will be in the range of 5 to 10. Equation (8.1) shows that for a given extraction ratio when sk/w is increased from 5 to 10, the change in the number of stages, n, is too small to matter. However, the actual value of n is itself low, pointing out the importance of a large k. With $s/w = 0.1$, $k = 50$, $sk/w = 5$, removal can change from 95 percent to 99.9 percent by increasing the extractor from about 2 theoretical stages to about 4. With excellent solvents, very high removal percentages are practical, whereas with poor solvents requiring many more theoretical stages they are not. This means that subsequent treatment for organic removal by biochemical oxidation, carbon adsorption, or whatever is required may not be necessary. The actual decision, however, is conditional not on how much phenol is extracted by the solvents but on how much of the total organic contamination is extracted by the solvent, and this is not known.

Solvent extraction is standard for all Lurgi gasification plants, where the proprietary Phenosolvan process is used. The scheme is considerably more complex than the idealized one discussed here, and descriptions of it may be found in Refs. 25 and 26.

Very little can be definitively said about the cost of solvent extraction. Each design must be the result of a more detailed economic evaluation than given here and must relate the solvent rate, the extraction ratio, and the number of extraction stages. The value of recovered phenol and other compounds is not known until they have been recovered in a test extraction with the chosen solvent and examined. Neither is it possible to state the quality of the effluent water without a test on each wastewater with each solvent. Even the choice of equipment requires investigation. Highly efficient liquid-liquid contacting equipment is available that can use low solvent to water ratios. But this equipment is expensive, and much simpler mixer-settlers may be optimal even though higher solvent to water ratios may be required.

8.4 Cooling Water

The last major component to be considered in a water treatment plant for a coal conversion complex is that part dealing with cooling water treatment. It has been emphasized throughout the book that cooling water makeup constitutes the largest single consumptive use of water in any coal conversion plant. Without the use of parallel wet/dry cooling systems to reduce consumption, the range of water evaporated for cooling in gasification plants producing 250×10^6 scf/day of pipeline gas was estimated to be between 700×10^3 and $1,400 \times 10^3$ lb/hr, or 2×10^6 and 4×10^6 gal/day. This is usually more than the dirty process condensate exported from the process, so cooling water must be drawn directly from the source of supply as shown in Fig. 8-2. A similar range of water evaporated for wet cooling was found for the coal liquefaction plants producing

50,000 barrels/day of liquid fuel and the 10,000 ton/day solvent refined coal plant (see Table 6.33).

Circulating cooling water is warm and well-oxygenated and is an ideal habitat for microbial growth. The water is seldom sterile when fed to the system and, in any case, receives a steady supply of air-borne growth. Untreated cooling systems are subject to fungal rot of the wooden parts of the tower, bacterial corrosion of iron and bacterial production of sulfide, growth of algae in the sun-lit portions of the tower, and suspended sloughed-off growth that can lodge in the system and block the flow. Biocidal chemicals must be added to control growth.

Not only may the makeup water contain silt, but the circulating water in its passage through the tower scrubs dust out of the air. Circulating water thus contains an ever-increasing amount of suspended matter, which will settle out in stagnant spots in the pipes and heat exchangers. This must also be prevented.

A wet cooling tower is an evaporator, and salts dissolved in the makeup water concentrate, often to the point of precipitation. The precipitate tends to adhere to heat transfer surfaces forming a hard scale. As in boilers, this must be prevented.

Finally, the well-oxygenated circulating water is very corrosive to heat transfer surfaces.

Treatment of cooling water is intended to prevent the problems of microbial growth, fouling, scaling, and corrosion. Although these classifications overlap and all the problems must be solved simultaneously, a discussion of each category provides an insight into the requirements for total control. At the present time the control of circulating cooling water is more of an empirical art than an engineering science.

In coal conversion plants with good water management, the treated process condensate should be used as makeup to the cooling system (see Fig. 8-2). This water will contain carbon, nitrogen, and, if biological treatment is used, phosphorus. All of these elements are important nutrients encouraging the growth of bacteria and algae. Except for the process water from the high temperature gasifiers, no matter how good the treatment of this water, it will probably also contain small amounts of phenol and other aromatic molecules.

The most common biocide for the control of microbial growth in circulating cooling water is chlorine. If chlorine is added, however, some of the aromatic molecules will be chlorinated, yielding chlorinated aromatics which, however dilute, are likely to be obnoxious if they escape. Chlorine should probably not be used as the cooling water biocide in synthetic fuel plants. Other biocides more expensive than chlorine are available.[26] With such biocides there is a strong incentive to minimize blowdown to minimize the loss of the biocide, but this can lead to other problems.

When the circulating cooling water has a high level of suspended solids, these tend to deposit in stagnant areas of the system, slowing the flow and possibly interfering with heat transfer. Uncontrolled growth will result in fouling. Other sources of suspended materials that can cause fouling are the air and the makeup water. Suspended solids introduced from these sources can be controlled by blowing down. The makeup water should usually be clarified if it is not clear. In very dusty atmospheres, where large blowdowns are not wanted, a side stream can be taken off the circulation loop, filtered, and returned. Suspending agents are regularly added to the circulating water to help prevent deposition of solids. Depending on the nature of the solids suspended in a given system, the rule of thumb is that operation can be successful provided the suspended solids are kept below 150 to 300 mg/l.

The absence of suspended solids and sloughed-off growth is not enough. As the circulating water is concentrated by evaporation, salts may precipitate. The salts which precipitate tend to be salts whose solubilities decrease as the temperature is raised. Thus precipitation is worst in the heat exchangers where the circulating water is heated. Precipitation often takes the form of hard scale adhering to the hot heat transfer surfaces and insulating the surface so that the rate of heat transfer is reduced. Among the precipitates that are troublesome are calcium carbonate, calcium sulfate, silica, magnesium silicates, calcium phosphate, and various other complex crystals.

Scale is customarily prevented by blowing down to avoid concentration. When the makeup is a natural water, the blowdown rate, dependent on the quality of the makeup water, is usually 30 to 50 percent of the makeup rate; that is, the water is blown down two to three times more concentrated than in the makeup. When the makeup water is partially or totally treated process condensate, the blowdown rate must be estimated from an analysis of the total makeup water. This amounts to deciding what concentrations of each pertinent species can safely be permitted in the circulating water.

The safe limits of concentration in circulating cooling water are empirically chosen from experience. A good and widely used set is given in Table 8.4.[27] The concentration products shown in the table are called solubility products. Usually they are the product of the concentrations of the positively and negatively charged constituents of the precipitate in solution. This product indicates the tendency to precipitate the solid phase.

Table 8.4
Limits for cooling tower circulating water composition.

	Conventional (Acidified to Low pH, No Precipitation)	Non-Conventional (High pH, Precipitation Dispersed into Suspension)
pH	6.5–7.5	7.5–8.5
Suspended Solids (mg/l)	200–400	300–400
Ca × CO$_3$ as (mg/l CaCO$_3$)^2	1,200	6,000
Carbonates (mg/l)	5	5
Bicarbonates (mg/l)	50–150	300–400
Silica (mg/l)	150	150–200
Mg × SiO$_2$ (mg/l)^2	35,000	60,000
Ca × SO$_4$ as (mg/l CaCO$_3$)^2	1.5 × 10^6 to 2.5 × 10^6	2.5 × 10^6 to 8 × 10^6

The headings "conventional" and "non-conventional" used in Table 8.4 require explanation. With the principal exception of calcium sulfate, the solubilities of scaling materials are greatly increased by decreasing the pH. A water in equilibrium with the carbon dioxide in the atmosphere is alkaline and has a pH about 8.3. In cooling towers equilibrium is only approached and its extent is empirically determined. In conventional cooling tower practice, a mineral acid is added to the cooling water to prevent precipitation and scale formation by reducing the carbon dioxide and pH. The concentration of the various scale forming components should then not be allowed to exceed the "conventional" limits shown in Table 8-4. This may be done by limiting the cycle of concentration in the cooling tower (see Section 4.1). It may be recalled that to minimize the ratio of makeup to evaporation rate and to reduce the blowdown rate, the cooling tower should be operated at the highest cycle of concentration. A limitation of operating with low pH is that the water is corrosive, requiring the addition of chemicals to inhibit corrosion. The best corrosion inhibiting chemical is soluble chromium,[28] which is toxic and persistent when released in the blowdown.

The need to avoid chromium corrosion inhibitors means that the circulating cooling water in coal conversion plants should probably be kept alkaline. Scale can be prevented by several water treatments. Some acid can be added to control carbonate scale, preferably just enough acid to hold the pH around 7.5. The makeup water can be softened by lime and soda to hold down the concentration of calcium, magnesium, and, to some extent, silica. A "non-conventional" procedure is to allow high concentrations of the precipitates with subsequent precipitation, but to add dispersant chemicals to cause the formation and suspension of very small crystals in the bulk of the water, rather than large, adhering crystals on the heat transfer surface. The suggested limits on concentration for this approach are given in the "non-conventional" column in Table 8.4. Usually such treatments make it possible to concentrate the makeup to the circulating system by tenfold to twentyfold so that the blowdown will be only 5 to 10 percent of the makeup. Without chromium the blowdown may be disposed by adding it to the ash sluicing systems.

In the water treatment scheme of Fig. 8-2 the cooling tower makeup is assumed to be drawn from a clean surface supply and treated process condensate. The principal cost for the cooling water treatment is for chemicals, and energy costs are negligible. It is not possible, however, to give a simple figure since the cost is highly dependent on whether the cooling tower is operated to a low or high cycle of concentration. Operation at a low cycle will incur waste of biocide and chemicals, while a high cycle will require makeup treatment. For the water supply noted, a range of \$0.15 to \$0.30/10^3 gal may be considered representative, though not for a design estimate that must be specific to the plant, location, and cooling tower scheme.

8.5 A Summary of Requirements

To place the various treatment procedures in perspective with regard to sequence and quantities handled, Fig. 8-9 shows a water treatment flow diagram for a Hygas plant to produce 250×10^6 scf/day of pipeline gas (see Table 5.11). This diagram is the same as Fig. 8-2 but with the details of the foul condensate treatment further broken down and with the flow quantities indicated. The climate is not arid and an adequate supply of fresh water is assumed so that combined wet/dry cooling of turbine condensers is not used. In the process condensate treatment chosen, phenol is not extracted but is destroyed by biological treatment. The water treatment may be considered representative where the coal is a low sulfur, surface-mined western coal. With the water taken to be a clean surface supply, the plant is illustrative for the coal fields of the Fort Union and Powder River regions. The design is neither wasteful of water nor one for minimum water consumption.

Cost and energy in water treatment requirements are much less well-defined than water quantities. Although the water treatment technologies considered are achievable ones, the experimental evidence for coal conversion process waters is

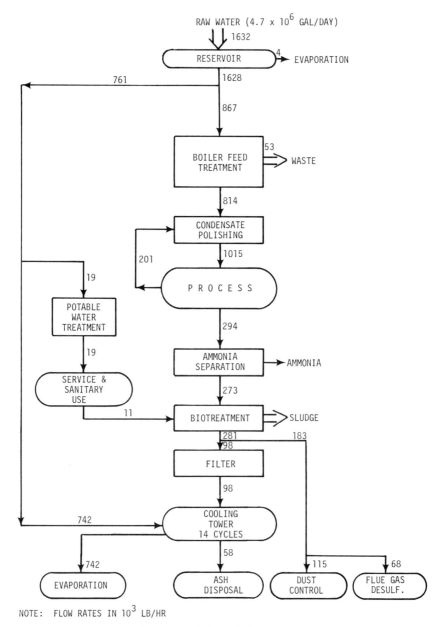

NOTE: FLOW RATES IN 10^3 LB/HR

Figure 8-9
Water treatment flow diagram for 250×10^6 scf/day Hygas plant located in the West with a clean and adequate surface water supply.

not available to fully assess them. For this reason, any designs and costs must be regarded with a greater degree of uncertainty than estimates of water quantity requirements.

A number of preliminary conceptual designs of water treatment plants have been carried out for the coal conversion processes examined in the book.[13] The costs derived for these designs were minimum ones not including engineering fees, site preparation work, buildings, and laboratories. For this reason, and because of the uncertainty indicated above, we assume that the actual values may be 50 percent greater than those derived. In this case the complete water treatment costs would be in the range of $0.02 to $0.10/10^6 Btu of product fuel. These costs take into account a credit for ammonia recovered. The water treatment charge is one that is not likely to exceed 5 percent of the sale price of the product fuel for any of the plants.

The energy required for the water treatment plants is very much controlled by the amount needed for ammonia separation. That amount, directly proportional to the rate of production of foul condensate, consumes more than 70 percent of the total water treatment energy. The total energy consumption has been estimated as a fraction of the product energy for the few conceptual designs of water treatment plants that have been made.[13] These estimates are more accurate than the treatment costs themselves. However, to take into account uncertainties in the amount of ammoniacal water produced and energy consumption not considered, we may assume that the actual consumption will be about 20 percent greater than the values derived in Ref. 13. In that case the total energy requirement for the water treatment plants will fall in the range of 0.7 to 2 percent of the product energy or 1 to 3 percent of the energy in the feed coal.

Finally, water will be consumed in the water treatment plants as regenerant wastes from the ion exchange units, as occluded water in treatment sludge, and by evaporation. Almost all of the consumed treatment water will be in the regenerant wastes. With a range of $1,000 \times 10^3$ to $1,500 \times 10^3$ lb/hr of water treated to boiler feed quality, and with a 5 to 15 percent volume of regenerant wastes, the water consumed may range from 50×10^3 to 225×10^3 lb/hr (100 to 450 gal/min).

Chapter 8. References

1. Spiegler, K. S. (ed.), *Principles of Desalination*. Academic Press, New York, 1966.

2. Sourirajan, S. (ed.), *Reverse Osmosis and Synthetic Membranes*. National Research Council of Canada, Ottawa, Canada, 1977.

3. Water Purification Associates, "Innovative Technologies for Water Pollution Abatement," Report No. NCWQ 75/13 (NTIS Catalog No. PB-247-390), National Commission on Water Quality, Washington, D.C., Dec. 1975.

4. Howe, E. D., *Fundamentals of Water Desalination*. Marcel Dekker, New York, 1974.

5. Babcock & Wilcox, *Steam/Its Generation and Use*. 38th Edition, Babcock & Wilcox, New York, 1975.

6. Simon, D. E., "Feedwater Quality in Modern Industrial Boilers—A Consensus of Proper Current Operating Practice," pp. 65–69, *Proc. 36th International Water Conference*, Engineers' Soc. of Western Pennsylvania, Pittsburgh, Nov. 1975.

7. Gehm, H. W. and Bregman, T. I., *Handbook of Water Resources and Pollution Control*. Van Nostrand Reinhold, New York, 1976.

8. Hammer, M. J., *Water and Waste-Water Technology*. Wiley, New York, 1975.

9. Strauss, S. D., "Water Treatment," *Power,* pp. S.1–S.24, June 1973.

10. Applebaum, S. B., *Demineralization by Ion Exchange*. Academic Press, New York, 1968.

11. Strauss, S. D., "Water Treatment," *Power,* pp. S.25–S.40, June 1974.

12. Gold, H., Goldstein, D. J., Probstein, R. F., Shen, J. S., and Yung, D., "Water Requirements for Steam-Electric Power Generation and Synthetic Fuel Plants in the Western United States," Report No. EPA-600/7-77-037, Environmental Protection Agency, Office of Energy, Minerals & Industry, Washington, D.C., May 1977.

13. Goldstein, D. J. and Yung, D., "Water Conservation and Pollution Control in Coal Conversion Processes," Report No. EPA-600/7-77-065, Environmental Protection Agency, Research Triangle Park, N.C., June 1977.

14. Milios, P., "Water Reuse at a Coal Gasification Plant," *Chemical Engineering Progress* **79** (6), 99–104, June 1975.

15. Melin, G. A., Niedzwiecki, J. L., and Goldstein, A. M., "Optimum Design of Sour Water Strippers," *Chemical Engineering Progress* **71** (6), 78–82, June 1975.

16. Treyball, R. E., *Mass-Transfer Operations*. McGraw-Hill, New York, 1968.

17. Dravo Corp., "Handbook of Gasifiers and Gas Treatment Systems," Report No. FE-1772-11, Energy Res. & Develop. Admin., Washington, D.C., Feb. 1976.

18. Metcalf & Eddy, Inc., *Wastewater Engineering*. McGraw-Hill, New York, 1972.

19. Kostenbader, P. D. and Flecksteiner, J. W., "Biological Oxidation of Coke Plant Weak Ammonia Liquor," *J. Water Pollution Control Federation* **41,** 199–207 (1969).

20. Scott, C. D., Hancher, C. W., Holladay, D. W., and Dinsmore, G. B., "A Tapered Fluidized-Bed Bioreactor for Treatment of Aqueous Effluents from Coal Conversion Processes," *Symp. Proc.: Environmental Aspects of Fuel Conversion Technology II,* pp. 233–240, Report No. EPA-600/2-76-149, Environmental Protection Agency, Research Triangle Park, N.C., June 1976.

21. U.S. Environmental Protection Agency, "Oxygen Activated Sludge Wastewater

Treatment Systems," Technology Transfer Seminar Publication, Environmental Protection Agency, Washington, D.C., Aug. 1973.

22. Bryan, E. H., "Two-Stage Biological Treatment Industrial Experience," pp. 136–153, *Proc. 11th Southern Municipal & Industrial Waste Conf.,* Univ. of North Carolina, Chapel Hill, N.C., 1962.

23. King, J. C., *Separation Processes.* McGraw-Hill, New York, 1971.

24. Shaw, H. and Magee, M. M., "Evaluation of Pollution Control in Fossil Fuel Conversion Processes. Gasification, Section 1: Lurgi Process," Report No. EPA-650/2-74-009c, Environmental Protection Agency, Research Triangle Park, N.C., July 1974.

25. El Paso Natural Gas Co., "Second Supplement to Application of El Paso Natural Gas Co. for a Certificate of Public Convenience and Necessity," Federal Power Commission Docket No. CP73-131, 1973.

26. Schultz, R. A., "Evolution of Non-Polluting Microbicides," Paper No. 131A, Cooling Tower Inst. Meeting, Cooling Tower Institute, Houston, Texas, Jan. 1974.

27. Crits, G. J. and Glover, G., "Cooling Blowdown in Cooling Towers," *Water and Wastes Engineering,* 45–52, April 1975.

28. Cone, C. S., "A Guide for Selection of Cooling Water Corrosion Inhibitors," *Materials Protection and Performance* **9,** pp. 32, 34, July 1970.

9
Resources and Water Requirements

9.1 Coal and Oil Shale

The locations of the principal coal and oil shale deposits in the conterminous United States are shown in Figs. 2-1 and 2-2. The largest coal deposits are in the Northern Great Plains and Rocky Mountain area encompassing the Powder River, Fort Union, and Four Corners regions, and in the Appalachian and Illinois Basins. High grade oil shale is found principally in the Green River Formation of the Rocky Mountain area.

For a given conversion efficiency and a fixed plant size (determined by the heating value of the product), the rate of coal mined is set by its heating value. For the three major coal ranks the following average heating values are used: bituminous, 13,000 Btu/lb; subbituminous, 9,800 Btu/lb; and lignite, 6,800 Btu/lb.[1] Table 9.1 shows the quantities of different rank coals that must be mined daily. These results extend those of Table 7.1 to the range of heating values encountered for all coal types, not just western coals. Also shown in Table 9.1 are the total recoverable reserves required to produce pipeline gas, fuel oil, and solvent refined coal by the specific processes noted. The "recoverable reserve" is the amount of coal actually mined or recovered as distinguished from the amount of coal present in the ground, or "coal reserve." The total recoverable coal reserve is about 50 percent of the total coal reserve for underground mining and about 80 percent for surface mining.[2,3]

Table 9.2 shows the approximate coal reserve required to support the coal needs of a single standard size mine-plant complex for the three major coal ranks and for both underground and surface mining. The coal reserve required for any one plant is but a fraction of a billion tons compared to the tens and hundreds of billions of tons of demonstrated coal reserves shown in Table 2.2 for the three principal coal basins of the United States.

Depending on the shale grade and the particular process, approximately 75,000 to 100,000 tons of high grade shale must be mined daily from an underground shale mine integrated with a shale oil plant, to produce 60,000 to 75,000 barrels/day of shale oil. This is the range of shale oil needed to produce 50,000 barrels/day of synthetic crude in a self-sufficient integrated plant (see Table 7.6). For one plant this means that a total recoverable reserve of 600×10^6 to 730×10^6 barrels of shale oil is needed assuming 325 days/yr production and a 30 year mine life. About 30 percent of the shale remains underground with conventional room-and-pillar mining techniques,[4] so that a total reserve of 860×10^6 to $1,040 \times 10^6$ barrels of shale oil is required for a plant producing 50,000 barrels/day of synthetic crude. This may be compared to identified reserves of about 570×10^9

Table 9.1

Coal mining rates and total recoverable reserve required for standard size plants producing synthetic fuels from three major coals.

	Pipeline Gas (Synthane) 250×10^6 scf/day	Fuel Oil (Synthoil) 50,000 barrels/day	Solvent Refined Coal (SRC) 10,000 tons/day
Daily Production Rate (tons/day)			
Bituminous*	15,800	16,100	17,300
Subbituminous*	20,900	21,400	22,300
Lignite*	30,100	30,800	33,100
Total Recoverable Reserve Required $(10^6$ tons)†			
Bituminous	154	157	169
Subbituminous	204	208	224
Lignite	294	300	323

*Heating values: bituminous, 13,800 Btu/lb; subbituminous, 9,800 Btu/lb; lignite, 6,800 Btu/lb.
†Based on 325 days/year production and a 30 year mine life.

to 620×10^9 barrels from high grade shale in the Green River Formation. Those reserves are more than sufficient to support a large scale oil shale industry.

Two of the most important factors in determining the recoverability of coal and oil shale are the bed or seam thickness and the bed depth or thickness of overburden. Overburden is generally any material that overlies the coal or oil shale deposits and is of little utility. Table 9.3 shows the categories of bed thickness used by the U.S. Geological Survey in estimating resources for the three ranks of coal considered.[2] These categories may differ from the recommended standards

Table 9.2

Total coal reserves required in millions of tons for a single standard size synthetic fuel mine-plant complex for underground and surface mining.

	Total Recoverable Reserve*	Total Coal Reserve	
		Underground Mining	Surface Mining
Bituminous	150	300	190
Subbituminous	200	400	250
Lignite	300	600	380

*Based on 325 days/year production and a 30 year mine life.

Table 9.3
Categories of bed thickness in inches used in calculating
resources of coal of different ranks.

	Thin	Intermediate	Thick
Bituminous	14–28	28–42	> 42
Subbituminous and Lignite	30–60	60–120	> 120

in a few states. Coal resource data is also categorized according to the thickness of overburden: 0 to 1,000 ft, 1,000 to 2,000 ft, and 2,000 to 3,000 ft. In some states, where the overburden is thin, other categories are used. The richest shale deposits are 30 ft or more thick with an overburden thickness of less than 1,500 ft.

Coal and shale resources are next considered by region.

Northern Great Plains and Rocky Mountain Area

The Northern Great Plains and Rocky Mountain area contain almost half of the nation's coal reserves, approximately 40 percent of which can be surface mined (see Table 2.2). Among the six coal regions in the Northern Great Plains, Fort Union and Powder River are the two largest. The largest coal deposits of the Rocky Mountain area are in the Four Corners region. The richest oil shale reserves in the United States are in the Green River Formation of the Rocky Mountain area.

The Fort Union region encompasses parts of eastern Montana, western North Dakota and northwestern South Dakota, with the latter containing minor amounts of strippable coal.[1,5] The region covers about 23,000 square miles (15×10^6 acres) and contains approximately 4.7×10^9 tons of strippable lignite. The coal seams are largely local in extent and are concentrated in at least six major coal fields. The average seam thickness is 16 to 20 ft in North Dakota and Montana with a maximum measured seam thickness of 85 ft in Montana. The major coal reserves are found within 200 ft of the surface. The lignite coal in this region has an ash content generally of less than 10 percent, a sulfur content of less than one percent, and a high moisture content that can exceed 40 percent. A typical coal analysis is given in Table 3.1. The major deposits have multiple seams with coal yields ranging from 14,000 to 40,000 tons/acre.

The Powder River region encompasses parts of southeastern Montana and northeastern Wyoming. The region covers about 23,000 square miles (15×10^6

acres) and contains approximately 15.4×10^9 tons of strippable low sulfur subbituminous C coal.[1,5] The seam thickness of the coal fields in Montana average about 28 ft, and Wyoming, which has the thickest coal seams in the United States, has an average seam thickness of 70 ft. The maximum seam thickness can be greater than 130 ft. The thickness of overburden is generally less than 200 ft. There are at least 14 major coal fields in the Powder River Basin; coal yields there range from 20,000 to 120,000 tons/acre. The ash content of the coal in this region is generally less than in the Fort Union region and averages less than 5 percent; the sulfur content is less than one percent. The moisture content averages about 20 percent. Typical coal analyses for this region are given in Tables 3.1 and 5.9.

The Four Corners region is comprised of parts of the states of New Mexico, Arizona, Utah, and Colorado. The largest quantities of surface minable reserves are in the San Juan Basin in northwestern New Mexico and in the Black Mesa Field in northeastern Arizona.[1,5] The San Juan Basin covers about 26,000 square miles (17×10^6 acres) and contains approximately 2.5×10^9 tons of strippable subbituminous B to A rank coal. The Navajo Field is the largest coal field in the San Juan Basin with an average coal seam thickness of about 11 ft and an overburden of less than 250 ft. The average ash content is about 20 percent with an average sulfur content of less than one percent and a moisture content of approximately 13 percent. Typical coal analyses for this field are given in Tables 3.1 and 6.10. The average yield is approximately 37,000 tons/acre. The Black Mesa Field covers about 3,200 square miles (2×10^6 acres) and contains surface coal reserves of approximately 582×10^6 tons. The coal rank varies from bituminous to subbituminous. The ash content of the coal ranges from 3 to 30 percent, the sulfur content from 0.4 to 2.3 percent, and the moisture content from 3 to 17 percent. The southwestern part of Utah contains large quantities of bituminous coal suitable for underground mining. This coal is found in the Kaiparowits Plateau.

The Green River Formation is in northwestern Colorado, northeastern Utah, and southwestern Wyoming. It contains high grade oil shale reserves with yields of between 25 and 65 gal/ton having a reserve equivalent to approximately 600×10^9 barrels of shale oil.[5-8] The richest and thickest deposits are found in Colorado in the Piceance Creek Basin in strata varying from 10 to 2,000 ft in thickness and with overburdens of up to 1,600 ft. The total reserves in this basin are estimated to be 480×10^9 barrels of oil. A typical ultimate analysis of organic matter in this oil shale is shown in Table 3.2. About 80×10^9 barrels of rich oil shale deposits are found in the Uinta Basin in Utah at depths to several thousands of feet below the surface. The oil shale deposits in Wyoming are found in the Green River Basin at overburden depths from 400 to 3,500 ft. The total reserves in this basin are estimated to be 30×10^9 barrels.

Appalachian Basin

The Appalachian Basin extends from northern Pennsylvania to northwestern Alabama and in between encompasses portions of Maryland, Virginia, West Virginia, Ohio, Tennessee, and Kentucky. Approximately 113×10^9 tons of coal, or 25 percent of the nation's demonstrated coal reserve base, is found in this region (see Table 2.2). Of the coal found there, 85 percent can be mined by underground methods and 15 percent by surface methods. Approximately 90 percent of the reserve in the Appalachian Basin is concentrated in Pennsylvania, Ohio, West Virginia, and eastern Kentucky, with 95×10^9 tons of underground reserves and 14×10^9 tons of surface reserves.[1,5] The region covers 70,000 square miles (45×10^6 acres) and contains primarily low volatile to high volatile bituminous coals. The coal seams are continuous over large areas and range in thickness from 2 to 6 ft with an average seam thickness of 4 ft. The ash content of the coal varies from 2 percent to as high as 50 percent with an average of 8 percent. The sulfur and the moisture content varies from less than one percent to greater than 3 percent, but is generally less than 3 percent. Typical coal analyses for Pittsburgh coals are shown in Tables 3.1 and 6.9.

Illinois Basin

Central and southern Illinois, southwestern Indiana, and western Kentucky comprise the Illinois Basin, which contains 89×10^9 tons of coal, or approximately 20 percent of the nation's coal reserves.[1,5] Approximately 18×10^9 tons can be recovered by surface mining. The seams are continuous over large areas, average about 3 to 4 ft thick, and lie at shallow depths of from 20 to 200 ft below the surface. The rank of the coal ranges from low volatile to high volatile C bituminous. Coal yields could exceed 19,000 tons/acre. The ash content of the coal ranges from 6 to 15 percent with an average moisture content of about 10 percent and with a relatively high sulfur content, generally in excess of 3 percent. Typical coal analyses for Illinois No. 6 C bituminous coal are shown in Tables 3.1 and 6.9.

9.2 Western Water Resources

Sufficient water is essential to the siting and operation of any synthetic fuel complex. However, reliability of the water supply is also a major consideration because of economic penalties that would be incurred in the shutdown of a plant due to a local water shortage. The availability of water is very specific to location, and the analysis of water supply for a given synthetic fuel mine-plant complex cannot be made without a detailed knowledge of the local supply and demand characteristics. With this limitation in mind, some general conclusions can be made concerning availability from regional considerations. Only those regions are considered in which the principal coal and oil shale deposits are located. It should be noted that most of the figures on availability given in this and the following section do not take into account seasonal or shorter term variations in flow.

Water supplies in the western states tend to be more variable over the year than in the eastern or central states. Annual rainfall is below the national average. Snow melt in the spring and hot dry summers result in most of the annual runoff occurring in a few months of the year—during the late spring and early summer. Runoff rates are very much lower for the remainder of the year. This has led to the construction of storage and conveyance facilities to provide adequate water supplies. Water in the West is used primarily for agricultural purposes, which is a high consumptive use. As a result, the smaller streams occasionally have years with several days of little or no flow and cannot meet the reliability criteria required for conversion plants without either large storage impoundments or the purchase of water rights.

In the West the adequacy of a water source is dependent on several factors, including the average quantity of water available at the intended source, the variability of the supply over time, the manner in which the water is appropriated, and the environmental and social implications involved in altering the hydrologic regime. The Appropriation Doctrine, defined in Section 2.5, is the code by which water is administered in the western states of concern. In this system, water rights are given priorities dependent on the seniority of the right and independently of the location of the water use with respect to its source. Generally, the only requirement regarding the use of water once a water right is confirmed is the need to put the water to "beneficial use," the definition of which is usually very loosely held. Water rights are considered to be property and can be bought and sold as such.

For large scale development in the states of interest, there are several other important factors influencing water availability. One of these pertains to the amount of water to be reserved for Federal lands, including the Indian reservations. Questions relating to this problem are now before the courts. Other constraints are the existing interstate river compacts, river basin compacts, and international treaties determining the allocation of water in river systems. Important among these are the Colorado River compacts and the Yellowstone River compact.

Any water supply strategy for synthetic fuel complexes must consider a number of sources and the consequences of development from among the alternatives. Included typically among the possible water sources are unused and unappropriated surface waters, groundwater, the purchase of surface and ground water rights, the development of adequate water storage facilities, and the possibilities of conjunctive use and water transfers over long distances. It should be noted, however, that the purchase of water rights presently used for other purposes is subject to many restrictions. The states are beginning to take a more active role in the supervision of these transfers.

Northern Great Plains and Rocky Mountain Area

In the Fort Union region, main-stem surface water supplies are generally adequate, although some of the smaller tributaries of the Missouri and Yellowstone rivers have low stream flows during part of the year. However, large scale synthetic fuel production will require considerable development of the water resources, including water transfer and storage facilities. The average precipitation is 15 in/yr, with most of the rainfall occurring during the spring and summer.[10] During the spring months the stream flows peak with the snow melts, while during the months of January and February the stream flows are at their lowest. In the Fort Union region, average open water surface evaporation is 36 in/yr in North Dakota and 38 in/yr in Montana.[10]

The principal rivers and tributaries in the Fort Union region are shown in Fig. 9-1. Table 9.4 gives the recorded surface runoff characteristics of some rivers at selected points in this region.[11,12] Runoff is an indicator of a region's water resources but it should not be used alone as a measure of water sufficiency. Taking a conservative estimate of average water use in a typical mine-plant complex to be 10 ft^3/sec (6.5 × 10^6 gal/day), Table 9.4 shows that the two largest rivers in the region, the Missouri and Yellowstone, have sufficient water to support a number of plants at the minimum discharge.

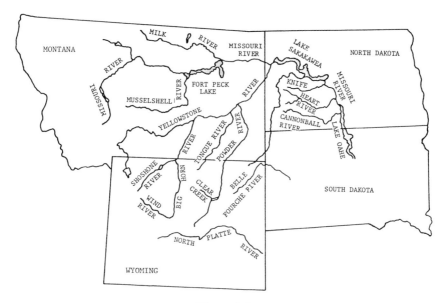

Figure 9-1
Major rivers in the Fort Union and Powder River regions.

Groundwater is available from alluvial and deep aquifers. The principal alluvial aquifers are located along the main rivers and their tributaries and have yields ranging up to 500 gal/min. Large amounts of water are stored in these alluvial aquifers. For example, almost 2×10^6 acre-ft (6.5×10^{11} gallons) are stored in the alluvial aquifers of the Yellowstone and Little Missouri rivers. Some of the deep aquifers are the coal seams. Yields from these aquifers in Montana and North Dakota range up to 400 gal/min. Groundwater is used as a water supply for many towns and rural homes.[5,13]

Table 9.4
Recorded surface runoff characteristics in cubic feet per second of rivers at selected points in the Fort Union region.

River and Location	Average Discharge	Maximum Discharge	Minimum Discharge
Missouri, near Culbertson, Montana	10,300	78,200	575
Yellowstone, near Sidney, Montana	13,030	159,000	470
Little Missouri, at Marmarth, North Dakota	343	45,000	0
Knife, near Hazen, North Dakota	183	35,300	0
Cannonball, near Breien, North Dakota	247	94,800	0

The principal use of water in the Fort Union region is for irrigation, although the major consumption of water is by evaporation from large reservoirs required for surface water regulation. Approximately 2.3×10^6 acres are irrigated annually above Lake Sakakawea, with plans for an additional 1.5×10^6 acres.[5] Estimates have been made of the water supply and the projected water needs of the Upper Missouri River Basin and the Yellowstone River subbasin, both of which include the Fort Union and Powder River regions. The main conclusion from a U.S. Department of the Interior study[14] is that there is sufficient water in the Upper Missouri River Basin to meet all projected needs and uses to the year 2000, including a high degree of energy development. Approximately 21.8×10^6 acre-ft/yr ($19,500 \times 10^6$ gal/day) of water is available with a projected need in the year 2000 of 4.0×10^6 acre-ft/yr ($3,600 \times 10^6$ gal/day). In the Yellowstone River Basin, a major tributary of the Missouri, approximately 8.8×10^6 acre-ft/yr ($7,900 \times 10^6$ gal/day) is available with a projected need of 1.3×10^6 acre-ft/yr ($1,200 \times 10^6$ gal/day).[14] The Northern Great Plains Research Project has indicated that the Missouri mainstream reservoirs could yield at least 2×10^6 acre-ft/yr ($1,800 \times 10^6$ gal/day) for industrial use in the Fort Union region.[15]

Surface waters in the Fort Union region have either sodium bicarbonate or sodium sulfate as the predominant constituent in their dissolved solids. The total dissolved solids range from 300 to 500 mg/l in the larger rivers and up to 2,000 mg/l in the tributaries, with relatively high average sediment concentrations exceeding 10,000 mg/l in some of the tributaries.[5] The groundwaters of the alluvial and deep aquifers are also of the sodium bicarbonate or sodium sulfate type, with total dissolved solids concentrations ranging from 500 to 6,000 mg/l.

Water supply in the Powder River region is similar to that in the Fort Union region, with major supplies provided by the Yellowstone and Bighorn rivers. During extended periods, however, the tributaries have negligible or zero flows. The rainfall and snowfall in this region average, respectively, about 14 and 36 in/yr, with rainfall heaviest in the spring and early summer.[10] During the early spring and summer, snow melt produces the largest annual streamflows, while lowest streamflows occur during late fall and winter. In the Powder River region the average annual open surface evaporation is 38 in/yr in Montana and 42 in/yr in Wyoming.[10]

The principal rivers and tributaries in the Powder River region are shown in Fig. 9-1. The recorded surface runoff characteristics of some rivers at selected points are given in Table 9.5.[11,16] The Yellowstone and Bighorn rivers have sufficient capacity under present conditions, even at minimum discharge, to

Table 9.5
Recorded surface runoff characteristics in cubic feet per second of rivers at selected points in the Powder River region.

River and Location	Average Discharge	Maximum Discharge	Minimum Discharge
Yellowstone, at Miles City, Montana	11,330	96,300	966
Yellowstone, at Billings, Montana	6,858	66,100	430
Tongue, at Miles City, Montana	423	13,300	0
Bighorn, at Bighorn, Montana*	3,851	26,200	275
Powder, at Arvada, Wyoming	272	100,000	0

*Regulated by storage facilities.

support a number of synthetic fuel plants, each having an average consumption of 10 ft³/sec (6.5×10^6 gal/day).

Several of the groundwater aquifers in the Powder River region have yields ranging from 5 to 500 gal/min. The Madison limestone aquifer, the largest in Montana and Wyoming, has produced well yields to 9,500 gal/min,[5,13] although the yields are highly variable with location.

The major consumptive use of surface water in this region is for irrigation. As previously noted, there is sufficient water in the Upper Missouri River Basin to meet future energy needs. In the Yellowstone River Basin, however, and particularly in the Powder River region, there is a problem of distributing the water from the source to the mine-plant complex because of the large distances involved. Estimates are that approximately 1×10^6 acre-ft/yr (890×10^6 gal/day) could be made available in this region to support energy development without the need to build new storage facilities.[15]

The average total dissolved solids range from less than 250 mg/l in the Yellowstone River to over 700 mg/l in the Bighorn River. The average annual sediment concentrations of some of the smaller tributaries can exceed 15,000 mg/l.[5]

The Four Corners region is quite arid. The San Juan and Little Colorado rivers shown in Fig. 9-2 are the principal rivers in the region. Precipitation ranges from 8 to 12 in/yr, with most of it coming mainly from heavy thunderstorms.[10] The peak flows of the San Juan River occur as a result of snow melt and thunderstorms, while the peak flow of the Little Colorado River is from direct surface runoff augmented by springs. The highest streamflows are in the spring and the lowest flows in early summer and winter. The annual average evaporation from

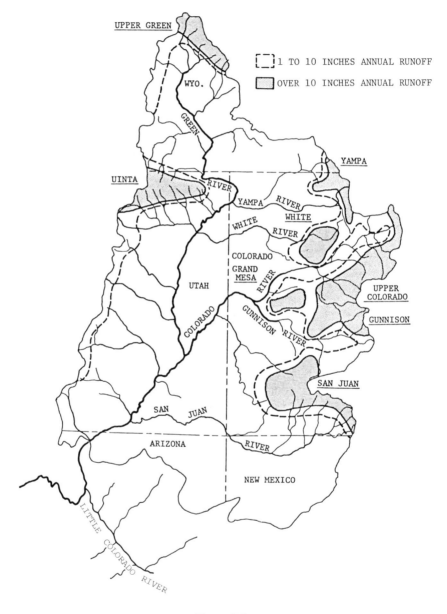

Figure 9-2
Major rivers and runoff producing areas in the Upper Colorado River Basin and the Little
Colorado River in the Lower Colorado River Basin.

open water surfaces is 46 in/yr in the San Juan Basin in New Mexico and 52 in/yr in the Black Mesa Field in Arizona.[10]

Recorded surface runoff characteristics of a number of the major rivers in the Four Corners region are shown in Table 9.6.[17,18] As in many parts of the West, the river flows are highly variable even with regulation of some of the rivers. The flows of the San Juan River are stabilized by the Navajo Reservoir with a capacity of over 1.7×10^6 acre-ft (0.55×10^6 gal).

In Black Mesa and the San Juan River Basin, there are a number of major aquifers. In Black Mesa, well yields range up to 2,000 gal/min, with many wells capable of 1,000 gal/min or more. In the San Juan River Basin well yields greater than 500 gal/min have been obtained.[5]

The San Juan River in New Mexico is part of the Upper Colorado River Basin. Under terms of the Colorado River Compact and the Upper Colorado River Basin Compact, New Mexico is allocated about 11 percent of the water available in the Upper Colorado River Basin after 50,000 acre-ft/yr is deducted for the Arizona allocation. The New Mexico entitlement ranges from 645,000 acre-ft/yr to 726,000 acre-ft/yr, according to different estimates.[19] Present water allocations total 566,000 acre-ft/yr and include a present use of 251,000 acre-ft/yr and a future use of 226,000 acre-ft/yr for a Navajo Indian Reservation project now under construction. In addition, 20,000 acre-ft/yr have been contracted for steam-electric power generation and 44,000 acre-ft/yr have been contracted from the Navajo Reservoir for coal gasification plants to be constructed in the future, for a total water allocation of 630,000 acre-ft/yr.[19] This leaves an additional 23,000 to 101,000 acre-ft/yr of water available for future energy development, or with the 44,000 acre-ft/yr contracted for gasification, a total of 67,000 to 145,000 acre-ft/yr (60×10^6 to 130×10^6 gal/day). There may also be the possibility of obtaining water presently included in the Navajo allocation. How-

Table 9.6

Recorded surface runoff characteristics in cubic feet per second of rivers at selected points in the Four Corners region.

River and Location	Average Discharge	Maximum Discharge	Minimum Discharge
Little Colorado, near Cameron, Arizona	201	24,900	0
San Juan, at Farmington, New Mexico	2,425	68,000	14
Animas, at Farmington, New Mexico	922	25,000	1
San Juan, near Carracas, Colorado	605	9,730	5

ever, the scarcity of surface water may still pose a serious constraint on the development of a synthetic fuels industry in the Four Corners region.

Water quality in the region varies not only because of varying flow conditions but because of industrial use and agricultural runoff. The San Juan River near Farmington contains moderate amounts of hardness (about 100 mg/l as $CaCO_3$), bicarbonates (about 120 mg/l as $CaCO_3$), and sulfate (about 120 mg/l as $CaCO_3$), with an average annual total dissolved solids concentration and suspended solids concentration, each of about 300 mg/l. The concentrations of the same quantities in the Little Colorado River are about twice that of the San Juan River. The total dissolved solids content of some of the groundwater in the San Juan Basin is greater than 1,000 mg/l, while in Black Mesa the dissolved solids range from about 1,000 mg/l to 25,000 mg/l.[5]

In the Green River Formation water is generally available from underground water sources as well as from the larger rivers of the Upper Colorado River Basin. Precipitation varies from 24 in/yr in the higher elevations of Colorado adjacent to the oil shale areas to 7 in/yr in Wyoming.[10] Snowfall accounts for most of the precipitation and falls during the winter and early spring. The average annual open water surface evaporation is about 36 in/yr in Colorado and about 42 in/yr in Wyoming.[10]

The principal rivers and tributaries in the Green River Formation are shown in Fig. 9-2, with the recorded surface runoff characteristics of some rivers at selected points in this region given in Table 9.7.[20] The highest runoffs occur in the spring and early summer as a result of snow melt, with smaller flows the rest of the year.

Significant quantities of groundwater are believed to be available only in the Piceance Creek Basin of Colorado. Estimates of the volume of water in storage in the deep aquifers in the Piceance Creek Basin range from 2.5×10^6 to 25×10^6 acre-ft (8.2×10^{11} to 82×10^{11} gal).[21,22] Groundwater is also available from shallower alluvial aquifers that are much smaller in areal extent than the deep aquifers. Recharge to the aquifers occurs mainly as a result of snow melt along the margins of the basin. Groundwater flows from the margins of the basin to the central part of the basin.[21] The surface water and groundwater systems are hydraulically connected so that if a large quantity of groundwater is withdrawn from an aquifer, flow in the neighboring streams could be decreased or possibly reduced to zero.

The rivers in the Green River Formation are part of the Upper Colorado River Basin. After 50,000 acre-ft/yr is given to Arizona, the remaining water is apportioned to the states in the Upper Colorado River Basin about as follows: Colorado

Table 9.7
Recorded surface runoff characteristics in cubic feet per second of rivers at selected points in the Green River Formation.

River and Location	Average Discharge	Maximum Discharge	Minimum Discharge
Colorado River, at Hot Sulphur Springs, Colorado	201	2,500	44
Colorado River, near Colorado-Utah State Line	5,345	33,000	1,570
Gunnison River, near Grand Junction, Colorado	2,072	12,000	500
Green River, near Green River, Wyoming	1,584	10,900	245
Green River, at Green River, Utah	5,811	29,500	1,180
Yampa River, at Steamboat Springs, Colorado	421	4,080	45
White River, near Meeker, Colorado	540	4,010	25

52 percent, Utah 23 percent, and Wyoming 14 percent.[19] Based on an assumed water supply of 5.8×10^6 acre-ft/yr, the following estimates have been made for the total water that could be made available for energy development: Colorado 90,000 acre-ft/yr (80×10^6 gal/day); Utah 128,000 acre-ft/yr (114×10^6 gal/day); and Wyoming 233,000 acre-ft/yr (208×10^6 gal/day).[5] Additional water would be available if there were transfers from existing uses.

An increase in salinity in the Colorado River is considered the most serious water quality problem.[9] The average salinity varies from less than 50 mg/l in the headwaters of the Colorado River to about 660 mg/l near Cisco, Utah, and to 610 mg/l at Lee Ferry, Arizona.[5,9] The dissolved solids concentration of groundwater in the Green River Formation ranges from 250 to over 60,000 mg/l, with the lower values occurring in the edges of the Piceance Basin, where hardness and bicarbonate ions predominate, and the higher values in the center of the basin, where sodium and bicarbonate ions predominate.

9.3 Eastern and Central Water Resources

In the eastern and central regions, variations in the surface runoff follow the variations in precipitation. Precipitation is way above the national average of 30 in/yr in these regions, and the normal distribution of surface water runoff does not vary as much as it does in the West. The major water use in the eastern and central states is municipal and industrial. Smaller streams generally have very few days of no flow and are more reliable than those in the West.

In the eastern and central regions, the use of surface flows is usually subject to the Riparian Doctrine, defined in Section 2.5. The owner of riparian land has the right to make use of the surface water in connection with the use of the riparian land as long as such use is "reasonable" with respect to others having a similar right. The Riparian Doctrine establishes an order of preference among various categories of users for determining a reasonable share; domestic users have the highest priority and industrial users a relatively low ranking.

If the use for energy development is less than about 5 percent of the low flow at a source, we will make the important assumption that the use is reasonable, since during normal flow periods the use would be an even smaller percentage of the flow. Another consideration in determining the adequacy of a water supply source involves the location of coal reserves at some distance away from the source. The non-riparian use restriction would suggest that either the coal from mining these reserves should be transported to conversion plants on major streams having riparian status, or conversion near the mine using non-riparian water should be considered only where the source has such an abundance of water that no legitimate riparian user can establish in a court of law that he has been injured by the non-riparian use.

Appalachian Basin

The water supplies of the major rivers of the Appalachian Basin shown in Fig. 9-3 are generally plentiful with total average streamflows of more than 150 billion gallons daily.[5] Surface water reservoirs within the region can store about 25 percent of the total average streamflow. Groundwater is generally abundant, but its availability varies throughout the basin. These water supplies are supported by ample rainfall and runoff. In the northern part of the basin the precipitation averages about 35 in/yr with more precipitation occurring in the late spring and summer.[10] The southern region receives an average of 55 in/yr of precipitation with most of the precipitation during winter and early spring. Surface water runoff averages 20 in/yr throughout the basin with some regions in the south averaging 30 to 40 in/yr. The evaporation from open water surfaces ranges from 28 in/yr in Pennsylvania to 42 in/yr in Alabama.[10]

For the purpose of determining the adequacy of a given water supply, we have chosen to compare the consumptive water requirements of a mine-plant complex with the expected low flows in the stream. In Table 9.8 are listed the recorded surface runoff characteristics of the major rivers at selected points in the Appalachian Basin. The 7 day, 20 year low flow is the minimum average flow over 7 consecutive days that is expected to occur with an average frequency of once in 20 years. This comparison is appropriate for plant sites having a useful life of

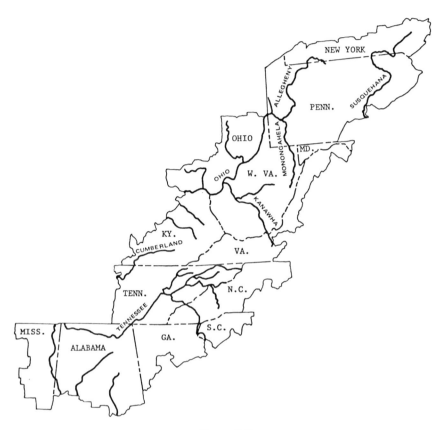

Figure 9-3
Major rivers of the Appalachian Basin.

about 20 years and holding ponds with a reserve capacity of a 7 day water supply. Stream low flow values were determined from U.S. Geological Survey or state records. Most of the major rivers in the Appalachian Basin are regulated with the result that higher low flows are achieved.

It is assumed that the average net water consumption of a typical mine-plant complex is approximately 10 ft³/sec, or about 6.5×10^6 gal/day. A stream is considered to have an adequate flow if its 7 day, 20 year low flow is at least 15 to 20 times the consumptive water use of a typical plant, or about 150 to 200 ft³/sec. From Table 9.8, within the Appalachian Basin, where coal is available, there are a number of large rivers that can provide a sufficient and reliable supply of water to support a large mine-plant coal conversion complex. This applies to all plant sites in the vicinity of the Ohio, Allegheny, Tennessee, and Kanawha-New

Table 9.8
Recorded surface runoff characteristics in cubic feet per second of rivers at selected points in the Appalachian Basin.

River and Location	Average Discharge	Historical Low Flow	7 Day, 20 Year Low Flow
Alabama			
Black Warrier, at Tuscaloosa	7,822	37	NA*
Tennessee, at Florence	51,610	105	NA*
Kentucky			
Ohio, at Evansville	133,900	NA*	15,400
Cumberland, at Cumberland Falls	3,199	4	12
Big Sandy, at Louisa	2,480	8	74
Ohio			
Tuscarawas, at Newcomerstown	2,453	170	215
Muskingum, at McConnelsville	7,247	218	565
Pennsylvania			
Allegheny, at Franklin	10,330	334	NA*
Conemaugh, at Seward	1,269	105	155
West Virginia			
Kanawha, at Charleston	14,480	2,360	1,750
Monongahela, at Greensboro	8,137	20	248
Cheat, at Rowlesburg	2,239	10	95

*NA = data not available because of extensive regulation.

rivers. However, the surface water supplies are much less reliable in the smaller streams, away from the major rivers. Regions generally found to have limited water supplies for energy development include the upper reaches of the Cumberland and Kentucky rivers in eastern Kentucky, the eastern Kentucky and adjacent West Virginia coal regions in the Big Sandy River Basin, and northern West Virginia and western Pennsylvania in the Monongahela River Basin, except those areas that can be supplied from the Allegheny or Ohio rivers. In these areas extreme low flows are practically zero, and a coal conversion complex could easily represent a significant portion of the seasonal low flow in many of these areas.

Groundwater can be an important conjunctive source of water in the Appalachian Basin.[5,13] Highly permeable aquifers with large yields occur in the valleys of the Allegheny and Ohio rivers. In the Muskingham River region in Ohio, well yields of greater than 500 gal/min have been obtained. Typical well yields in the valley of the Monongahela River in West Virginia are 25 gal/min for consolidated aquifers, but the deep sandstone aquifers can have yields as high as 400 gal/min. In northern Alabama only a few aquifers provide high yields, with the majority ranging in yields from 25 to 100 gal/min. In eastern Kentucky low yields ranging from 10 to 25 gal/min are generally obtained.

Unlike the western regions, water for irrigation is very small because of above average rainfall and the comparatively small amount of farmland. Municipal and industrial use accounts for approximately $11,000 \times 10^6$ gal/day or about 7 percent of the total average stream flow.[5] Large quantities of water withdrawals are required for steam–electric power generation, but the consumptive use is relatively small, amounting to approximately $1,000 \times 10^6$ gal/day.

The water quality in most of the rivers is good, with dissolved solids concentrations of less than 300 mg/l. Some of the river waters have deteriorated because of municipal and industrial pollution and acid mine drainage. The quality of the groundwaters in this region is in general satisfactory for most domestic uses. Some underground waters are high in calcium bicarbonate.

Illinois Basin

The situation in the Illinois Basin is similar to that in the Appalachian Basin with respect to water supply. Both surface water and groundwater are abundant in the major coal producing areas of the Illinois Basin and are supported by ample rainfall and surface runoff.[23] The average precipitation ranges from 35 to 40 in/yr in central Illinois to about 48 in/yr in western Kentucky.[10] In the northern part of the basin most of the precipitation occurs in the spring, while in the southern part the highest precipitation occurs in mid-winter and early spring. The average annual surface runoff ranges from 8 in/yr in the northern region to 15 in/yr in the southern region, with the highest runoff occurring at the same time as the highest precipitation. The annual average evaporation from an open water surface is 33 in/yr in Illinois and 36 in/yr in western Kentucky.[10]

The rivers of the Illinois Basin are shown in Fig. 9-4 and the recorded surface runoff characteristics of the major rivers at selected points are shown in Table 9.9. The Ohio and Mississippi rivers have sufficient and reliable water supplies to support a large mine-plant coal conversion complex. The lower sections of the Kaskaskia, Illinois, and Wabash rivers in Illinois,[23] the Wabash and White rivers in Indiana, and the Green River in Kentucky also have reliable supplies.

Groundwater is available throughout the region from alluvial and bedrock aquifers. Well yields between 75 to 1,000 gal/min have been obtained from the

Figure 9-4
Major rivers of the Illinois Basin.

alluvial aquifers along the major rivers in this basin. Bedrock aquifers found in Illinois can yield up to 500 gal/min.

The Illinois Basin is located in both the Upper Mississippi and Ohio river basins, which have a daily surface outflow of about 234,000 × 10⁶ gal/day.[5,24] Municipal and industrial use accounts for approximately 20,000 × 10⁶ gal/day with 90 percent used for industrial purposes. Water use for irrigation is negligible.

The average dissolved solids content of the major rivers in the region range from 200 to 700 mg/l but can go higher during periods of low flow. The surface waters in the Illinois Basin are generally hard.[5] The dissolved solids content of the groundwaters generally do not exceed 500 mg/l. High levels of iron, hardness, and sulfates occur in many of the groundwaters.

Table 9.9
Recorded surface runoff characteristics in cubic feet per second of rivers at selected points in the Illinois Basin.

River and Location	Average Discharge	Historical Low Flow	7 Day, 20 Year Low Flow
Illinois			
Illinois, at Kingston Mines	14,529	1,810	NA*
Kaskaskia, at Shelbyville	788	0	NA*
Mississippi, at Keokuk, Iowa	62,570	5,000	6,500
Indiana			
White, at Petersburg	11,540	573	610
Wabash, at Riverton	11,600	858	350
Ohio, at Evansville	113,700	NA*	13,500
Kentucky			
Green, at Calhoun	10,960	280	NA*

*NA = data not available because of extensive regulation.

9.4 Water Consumption and Residual Totals

In this section the net water consumption and wet solid residuals generated are summarized for standard size mine-plant complexes assumed to be located in each of the five principal coal regions and in the principal oil shale region of the United States. In order that the results may be scaled to plant sizes different than those considered in the book, water consumption and residuals generated are also expressed per unit of heating value in the product fuel. All of the summaries are based on specific results given in earlier chapters or on the methodology outlined there.

Several synthetic fuel technologies are compared: coal gasification to convert coal to pipeline gas; coal liquefaction to convert coal to low sulfur fuel oil; clean coal to produce a de-ashed, low sulfur solvent refined coal; and oil shale to produce synthetic crude. Table 9.10 lists these technologies and the processes chosen to illustrate them, together with a summary of the product fuel output and heating value for the standard size plants examined.

Table 3.1 gives the analyses of the coals used. Each coal is representative of one in each of the five principal coal mining regions of the country. The coal is assumed to be surface mined, except in the Appalachian region where we have

Table 9.10
Product fuel output of standard size synthetic fuel plants.

Technology and Conversion Process	Product	Output	Product Heating Value (10^{11} Btu/day)
Coal Gasification			
Synthane	Pipeline Gas	250×10^6 scf/day	2.4
Hygas			
Lurgi			
Coal Liquefaction			
Synthoil	Fuel Oil	50,000 barrels/day	3.1
Clean Coal			
SRC	Solvent Refined Coal	10,000 tons/day	3.2
Oil Shale			
Paraho Direct	Synthetic Crude	50,000 barrels/day	2.9
Paraho Indirect			
TOSCO II			

also assumed it is underground mined in order to estimate the differences in quantities of water required and residuals generated between the two methods. A typical analysis for a Green River oil shale is given in Table 3.2.

The daily coal mining rates are shown in Table 9.11 for the coal conversion plants utilizing the three different coal types. The rates include any coal fed to the boiler as well as to the gasifier or reactor. The shale grade and mining rates are given in Table 7.6.

Table 9.11
Coal mining rates in thousand tons per day for standard size plants producing synthetic fuels.

Process	Coal Rank		
	Lignite	Subbituminous	Bituminous
Synthane	30.5	25.5–25.9	16.6–17.8
Hygas	30.1	21.3–24.4	15.8–18.1
Lurgi	33.0	29.4–31.7	16.2–18.8
Synthoil	31.6	26.5–27.5	16.0–19.6
SRC	33.7	28.3–30.6	18.3–22.4

Net Water Consumed

Figure 9-5 shows the variation with region of the net water consumption for the coal gasification processes, and Fig. 9-6 the variation for the coal liquefaction and clean coal plants. All of the coal conversion examples correspond to the case of high wet cooling, defined below. For the oil shale plants only the high grade oil shale region and only the case of intermediate wet cooling defined below are considered. Reuse of the dirty process condensate is always assumed. Except for the Four Corners region, differences in water consumption between locations for a given coal conversion process are relatively small, being no more than 15 percent for both the Lurgi and the Synthoil processes. The higher consumption for the Four Corners region is the result of the larger amount of water needed for dust control and ash handling of coal with a high ash content. Water is also required for revegetation, unnecessary in the other locations. Water consumed in water treatment is not included in the total consumption but is estimated to be less

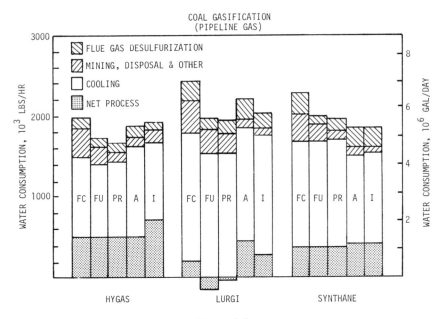

Figure 9-5
Net water consumption for standard size coal gasification plants (FC, Four Corners region; FU, Fort Union region; PR, Powder River region; A, Appalachian Basin; I, Illinois Basin).

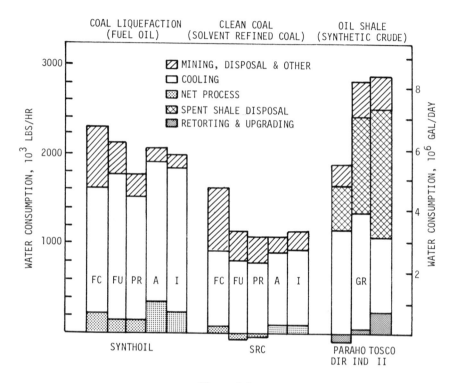

Figure 9-6

Net water consumption for standard size coal liquefaction, clean coal, and oil shale plants
(FC, Four Corners region; FU, Fort Union region; PR, Powder River region; A, Appalachian Basin; I, Illinois Basin; GR, Green River Formation).

than 5 percent of it. The effect of the coal conversion process on the net water requirements in the Four Corners region is shown in Fig. 9-7.

Since the products and outputs of the Synthane, Hygas, and Lurgi coal gasification plants are the same, these three facilities may be compared directly. A Lurgi plant uses from 25 to 45 percent more water than does a Hygas one at the same site, and this represents the largest difference between the processes. This difference, due to differences in estimated overall efficiency, is even larger than it appears on the surface, because the Lurgi process accepts wet coal and the resulting dirty condensate has been subtracted from the process requirement. However, this apparent lowering of consumption in the Lurgi process is not without cost since the foul condensate must be cleaned before it is reused in the

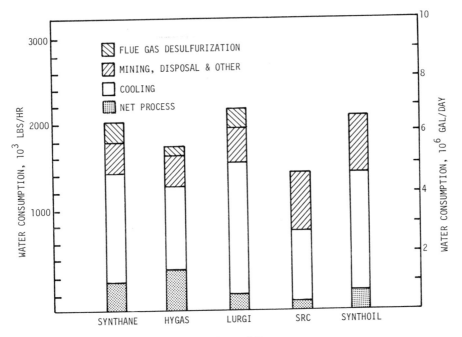

Figure 9-7
Net water consumption for standard size coal conversion plants in the Four Corners region.

plant. Disposal of the foul condensate is not practical, requiring cleaning before disposal to meet environmental regulations.

The products of the Synthoil and oil shale plants are roughly the same, so they too may be compared. The water consumption for the Synthoil process is comparable with the Paraho Direct heated process. However, the water consumed in the other two oil shale processes (Paraho Indirect and TOSCO II) is 60 percent higher due mainly to the larger water requirements for spent shale disposal and revegetation.

One means of comparing the water requirements for different processes and products is to normalize the net water consumption with respect to the heating value in the product fuel. Normalized water requirements in gal/10^6 Btu are listed in Table 9.12, with no distinction between sites and the different coal types, but with overall ranges given instead. The highest values correspond to the Four Corners region.

Table 9.12
Net water consumed in gallons per million Btu of heating value in the product fuel.

	Minimum Practical Wet Cooling	Intermediate Wet Cooling	High Wet Cooling
Coal Gasification			
Synthane	13–16	14–18	22–28
Hygas	14–18	15–19	20–24
Lurgi	12–19	18–26	21–29
Coal Liquefaction			
Synthoil	11–16	12–16	17–20
Clean Coal			
SRC	5–10	5–11	10–15
Oil Shale			
Paraho Direct	—	19	—
Paraho Indirect	—	31	—
TOSCO II	—	31	—

Three separate situations are indicated in Table 9.12, representing the use of different amounts of wet evaporative cooling for the turbine condensers and interstage coolers on the gas compressors. In arid regions or in regions where water is expensive, combined wet/dry cooling is assumed to be used for both the turbine condensers and interstage coolers. In this case "minimum practical wet cooling" is assumed, and wet cooling is used to handle 10 percent of the cooling load on the turbine condensers and 50 percent of the load on the interstage coolers. For the Lurgi process, a detailed thermal balance is not available, and wet cooling for this case is assumed to be used to dispose of 18 percent of the total unrecovered heat. This value is the one estimated for other pipeline gas plants in the same areas. In regions where water is available but is moderately expensive, it is assumed that wet cooling handles 10 percent of the cooling load of the turbine condensers and all of the load of the interstage coolers. This is the case of "intermediate wet cooling." For the Lurgi process it is assumed that in this case 30 percent of the unrecovered heat is disposed by wet cooling. For the oil shale processes, the percentage of unrecovered heat removed by wet cooling is listed in Table 6.33. These percentages are considered to correspond to intermediate wet cooling. The last case of "high wet cooling" is appropriate for those regions where water is plentiful and inexpensive. Here, the loads on both the turbine condensers and interstage coolers are taken to be all wet cooled, and 36 percent of the unrecovered heat in the Lurgi process is assumed to be disposed by wet cooling. All the results presented in Figs. 9-5 and 9-6 correspond to this case.

The maximum difference in water consumption between high wet cooling and minimum practical wet cooling at a given site is approximately 12 gal/10^6 Btu for the Synthane process. The minimum practical cooling will cost $0.018/$10^6$ Btu more than high wet cooling and will be used if water is relatively expensive, that is about $1.5/$10^3$ gal or more. This cost is not likely to exceed one percent of the sale price of the product fuel and represents a small price to significantly reduce water consumption.

Table 9.12 shows that solvent refined coal production has the lowest water requirement for any given cooling design. The next lowest requirement is for the conversion of coal to oil by the Synthoil process. The largest coal conversion water requirements are to produce pipeline gas, with the consumption about the same for all of the processes. To convert oil shale to synthetic crude by the Paraho Direct process requires about the same water as the conversion of coal to pipeline gas. The Paraho Indirect and TOSCO II processes for the conversion of oil shale have the largest water requirement.

The net water consumed for the case of maximum wet cooling is about twice that for the case of minimum practical wet cooling. Even more dramatic is the factor of six between the low value for the Solvent Refined Coal process and the high one for the Lurgi process. The point to be made is that the amount of water consumed to manufacture synthetic fuel depends importantly on the product and the cooling design.

In the Appalachian Basin estimates were also made for underground mining without coal washing. It was found that the additional water needed for mine dust control is not more than 3 percent of the total mine-plant water requirements when surface mining is used.

Table 9.13 shows the net water consumed in million gallons per day for the standard size plants with the various wet cooling options described. For the case of intermediate wet cooling, the water requirements at the low end of the spread range from 1.7 × 10^6 gal/day for the Solvent Refined Coal process to 4.4 × 10^6 gal/day for the Lurgi process to 8.9 × 10^6 gal/day for the Paraho Indirect and TOSCO II processes. The low value of 1.5 × 10^6 gal/day for the Solvent Refined Coal process compared with the high value of 6.9 × 10^6 gal/day for the Lurgi process represents a difference of a factor of about five in absolute water requirement.

To illustrate how the water is consumed, a breakdown of the consumption by use is also shown in Figs. 9-5 to 9-7. To simplify the graphical presentation, only

Table 9.13
Net water consumed in million gallons per day for standard size synthetic fuel plants.

	Minimum Practical Wet Cooling	Intermediate Wet Cooling	High Wet Cooling
Coal Gasification			
Synthane	3.0–3.8	3.4–4.2	5.2–6.5
Hygas	3.4–4.3	3.6–4.5	4.8–5.8
Lurgi	2.9–4.6	4.4–6.1	5.1–6.9
Coal Liquefaction			
Synthoil	3.3–4.8	3.6–5.1	5.8–6.7
Clean Coal			
SRC	1.5–3.3	1.7–3.5	3.1–4.8
Oil Shale			
Paraho Direct	—	5.5	—
Paraho Indirect	—	8.9	—
TOSCO II	—	8.9	—

four water use categories are presented for each coal conversion process at each site: net process water based on reuse of all condensates; cooling water; flue gas desulfurization water, if necessary; and water for mining, dust control, solids disposal, and other uses. For oil shale, the water use categories have been subdivided somewhat differently to reflect the large requirement for spent shale disposal: net process water for retorting and upgrading; cooling water; water to dispose of the spent shale and for revegetation; and water for dust control, mining, and other uses.

For the high wet cooling option depicted in Figs. 9-5 to 9-7, the largest single consumptive use of water is for cooling in all the facilities except the TOSCO II oil shale plant. For the same energy input in the raw fuel and for the same fraction of unrecovered heat dissipated by wet cooling, the higher the overall conversion efficiency the lower the cooling water consumption. A comparison of cooling requirements is shown in Table 9.14 for the three cooling options considered in this section. By using combined wet/dry cooling of the turbine condensers and interstage coolers, approximately one half to two thirds of the cooling water can be saved.

Figures 9-5 and 9-6 show that the process water requirements for the Synthane and, to some extent, the Hygas plants are independent of site, while the consumption by the Lurgi and Synthoil processes are site-dependent. Process requirements for the Synthane plants are less than for the Hygas plants because the

Table 9.14
Cooling water requirements in gallons per million Btu of heating value in the product fuel.

	Minimum Practical Wet Cooling	Intermediate Wet Cooling	High Wet Cooling
Coal Gasification			
Synthane	5.9–6.5	6.9–8.0	12.9–16.4
Hygas	5.3–5.5	6.0–6.1	11.1–11.4
Lurgi	8.6–9.6	14.2–15.9	17.3–19.3
Coal Liquefaction			
Synthoil	7.0–8.2	7.9–9.1	12.9–15.0
Clean Coal			
SRC	2.7–2.8	3.2–3.3	7.3–7.5
Oil Shale			
Paraho Direct	—	11.6	—
Paraho Indirect	—	13.3	—
TOSCO II	—	8.5	—

Synthane process makes char and passes more coal through the gasifier. This makes more hydrogen available from coal. For the Lurgi process the most important factor in the variation in process water consumption is that the plant accepts wet coal and the coal moisture overshadows the requirement, though the resulting dirty process condensate must be treated. For the Synthoil and Solvent Refined Coal plants, the net process water requirements are a function of the oxygen content, with the highest requirement in the Appalachian Basin corresponding to the lowest oxygen content in the coal and the lower requirements in North Dakota and Wyoming corresponding to the higher oxygen contents in the coal. However, when hydrogen is produced from very moist coals, principally lignite, without predrying the coal, the net process water consumption will be less than that indicated by the oxygen content. The process water consumption or production in the oil shale plants relate directly to the amount of water produced in the retort itself, as shown in Table 6.19.

The largest single factor in the water requirement for flue gas scrubbing is the moisture content of the coal or char fed to the boilers. For this reason the flue gas requirement is greatest for the coals from the Appalachian and Four Corners regions, which are relatively low in moisture. In all Synthane plants dry char is fed to the boiler making the scrubbing water requirements high. Coal is not fed to the boilers in the Solvent Refined Coal and Synthoil designs considered.

The water required for ash disposal, dust control, and other needs cannot be readily generalized because of the many competing factors. However, the water requirements for the Four Corners region are higher than for the other four regions because of the high ash coal and the revegetation requirement. The water requirements for the disposal of spent shale and subsequent revegetation differ considerably between processes, depending on the operator's assumption about the amount of water necessary to properly dispose of the spent shale. In the TOSCO II plant it is assumed that the spent shale is both compacted and cemented with water, while in the Paraho designs the spent shale is simply compacted dry. The water consumption for the Paraho design is mainly for revegetation, whereas in the TOSCO II design it is in large part for compaction.

Residuals Generated

Every synthetic fuel plant generates solid residuals, most of which will be disposed wet with occluded water that is lost and chargeable to the process. In the coal conversion plants these residuals are principally the coal ash and, where flue gas scrubbing is used, the flue gas desulfurization sludge. In the oil shale plants

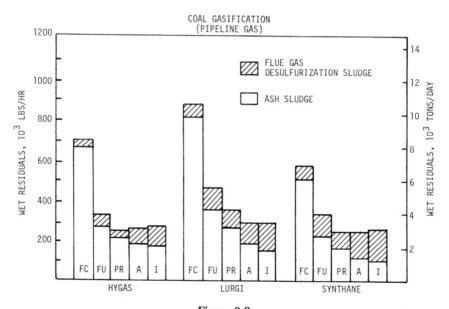

Figure 9-8
Amount of wet residuals generated by standard size coal gasification plants (FC, Four Corners region; FU, Fort Union region; PR, Powder River region; A, Appalachian Basin; I, Illinois Basin).

the principal residual is the spent shale. Figures 9-8 and 9-9 present the amount of main wet residuals, including occluded water, generated by each standard size plant in each region. Table 9.15 gives the range of wet residuals generated by each process and the range normalized with respect to the heating value in the product fuel. Not included are the residuals generated in water treatment since they are generally less than one percent (and in some extreme cases of very hard waters up to 5 percent) of the flue gas desulfurization and ash sludges.

The residuals generated are highest for the Four Corners region because of the high ash content of the coal compared to the other regions. The lower values of the total residuals generated per million Btu in the product fuel for all of the conversion processes are in a narrow band ranging from 23 lbs/10^6 Btu for the Hygas process to 29 lbs/10^6 Btu for the Lurgi process. The predominant solid waste from these processes is ash, and the variation in range relates to the fraction of waste that is bottom ash, fly ash, or flue gas desulfurization sludge;

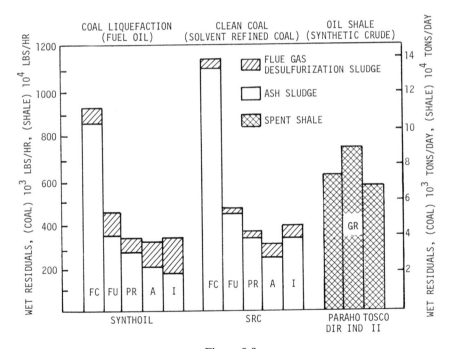

Figure 9-9
Amount of wet residuals generated by standard size coal liquefaction, clean coal, and oil shale plants (FC, Four Corners region; FU, Fort Union region; PR, Powder River region; A, Appalachian Basin; I, Illinois Basin; GR, Green River Formation).

Table 9.15
Wet residuals generated by standard size synthetic
fuel plants.

	10^3 tons/day	lb/10^6 Btu*
Coal Gasification		
Synthane	2.8–7.0	24–59
Hygas	2.8–8.5	23–70
Lurgi	3.5–11	29–88
Coal Liquefaction		
Synthoil	3.7–11	24–72
Clean Coal		
SRC	4.0–14	25–86
Oil Shale		
Paraho Direct	76	520
Paraho Indirect	91	630
TOSCO II	68	470

*In the product fuel.

each requires somewhat different quantities of water for disposal. Outstripping all of the coal conversion residuals by an order of magnitude are those from oil shale processing where the primary residual is the spent shale. Comparing the coal conversion and Paraho Indirect facilities, the quantity of shale residuals is approximately 30 times those from the coal conversion plants on a normalized basis (lb/10^6 Btu) for all regions except Four Corners. For the Four Corners region the quantity of shale residuals is a factor of 7 to 14 times those of the coal conversion plants. For the Lurgi plants, 3.5×10^3 to 11×10^3 tons/day of wet residuals are generated, with the highest value corresponding to the Four Corners region. This may be compared to the range of 68×10^3 to 91×10^3 tons/day for the oil shale facilities.

9.5 Water Requirements for Synthetic Fuel Production

The results of the last three sections are brought together here to determine if in terms of water requirements and availability, synthetic fuel production can be supported in each of the principal coal or oil shale regions of the United States and to determine at what level that production can be supported. Other environmental limitations beyond those imposed by water availability have not been considered. Socioeconomic limitations have also not been considered. It should

Table 9.16
Number of standard size plants required to produce 1×10^6 barrels/day of synthetic crude
or its equivalent of 5.8×10^{12} Btu/day.*

Conversion Technology	Product	Unit Output	Number of Standard Size Plants
Coal gasification	Pipeline Gas	250×10^6 scf/day	24
Coal liquefaction	Fuel Oil	50,000 barrels/day	19
Clean coal	Solvent Refined Coal	10,000 tons/day	18
Oil shale	Synthetic Crude	50,000 barrels/day	20

*See Table 9.10 for product heating value.

be kept in mind that the availability of water is site-dependent, and although a region may have sufficient water on the average to support energy development, it may not have sufficient water in localized areas.

For purposes of illustration, we consider a synthetic fuels industry producing 1×10^6 barrels/day of synthetic crude, or its equivalent in other fuels of 5.8×10^{12} Btu/day, in each of the five principal coal bearing regions and in the principal oil shale region of the country for a total production of 6×10^6 barrels/day. As a relative measure, in 1977 crude oil was imported at about the rate of 6×10^6 barrels/day and distilled products at about 2×10^6 barrels/day. Table 9.16 lists the number of standard size plants required to produce 5.8×10^{12} Btu/day for the conversion technology and product output indicated. The range is from 18 clean coal plants producing 10,000 tons/day of solvent refined coal to 24 coal gasification plants producing 250×10^6 scf/day of pipeline gas.

For each of the coal conversion technologies producing 5.8×10^{12} Btu/day, the total water requirements are shown in Table 9.17 for the cases of high wet cooling and minimum practical wet cooling. For coal gasification, the low and high ends of the range were derived using averages of the low and high values in

Table 9.17
Water consumption in million gallons per day to convert
coal to 1×10^6 barrels/day of synthetic crude or its
equivalent of 5.8×10^{12} Btu/day.

	Minimum Practical Wet Cooling	High Wet Cooling
Coal gasification	75–104	122–157
Coal liquefaction	64–93	99–116
Clean coal	29–58	58–87

Table 9.12 for the three gasification processes. The coal liquefaction and clean coal values were derived from the values given in Table 9.12 for the Synthoil and Solvent Refined Coal plants. The high end of the range for all the coal conversion plants correspond to the requirements for the Four Corners region.

In Table 9.18 consumption is presented for each of the five coal bearing regions as a fraction of the estimated average annual values of available surface water. Substantially higher fractional consumptions could occur during dry months and occasional dry years. The range at a given site corresponds to the difference in the amount of wet cooling used. For the Appalachian and Illinois basins, the water available corresponds to the total daily surface runoff from both the Upper Mississippi and Ohio river basins. There is more than enough water to support production of 1×10^6 barrels/day of synthetic crude oil or its equivalent in either the Appalachian or Illinois basins. In the Powder River and Fort Union regions the water consumed ranges from about 1.5 to 14 percent of the average annual water supply in the region, depending on the particular process and degree of wet cooling selected. The high end of the range is most probably an excessive demand and might not be acceptable. The low end of the range could be considered a reasonable level for most of the areas, assuming suitably formulated water use plans. These might include the purchase of existing water rights, the development of surface and ground water facilities, and cooperative use practices. The Four Corners region is certainly not able to support the production levels indicated.

Oil shale conversion is not included in Tables 9.17 or 9.18, because only the

Table 9.18
Water consumption as a percentage of available surface water to convert coal to 1×10^6 barrels/day of synthetic crude or its equivalent of 5.8×10^{12} Btu/day.

Region	Available Surface Water† (10^6 gal/day)	Percent of Available Surface Water		
		Coal Gasification	Coal Liquefaction	Clean Coal
Powder River	890	8.4–13.7	7.2–11	3.3–6.5
Fort Union	1,790	4.2–6.8	3.6–5.5	1.6–3.2
Four Corners	60	*	*	98–*
Appalachian and Illinois Basins**	230,000	0.066–0.106	0.056–0.086	0.026–0.050

†Average annual values; supporting assumptions given in Sections 9.2 and 9.3.
*Greater than available water supply by a factor of about 1.5 to 2.5.
**Two million barrels/day of synthetic crude or its equivalent of 11.6×10^{12} Btu/day.

Table 9.19
Water consumption to convert oil shale to 1×10^6
barrels/day of synthetic crude.

Available Surface Water† (10^6 gal/day)	Consumption	
	10^6 gal/day	Percent of Available Water*
210	110–180	52–86

†Average annual value; supporting assumptions given in Section 9.2.
*Green River Formation (Colorado and Utah).

requirements for one cooling option were given. An equally, and in some instances even more important, factor affecting consumption in surface oil shale processing is the method of spent shale disposal. Table 9.19 shows for the one cooling option considered the range of water consumption for a 1×10^6 barrel/day synthetic crude industry. The ranges given correspond to the different retorting procedures and disposal methods summarized in Table 9.12. Because the largest quantity of high grade oil shale is in Colorado and Utah, water availability for the Green River Formation was estimated only for these states. It would appear from the figures in Table 9.19 that a production level of 1×10^6 barrels/day of synthetic crude from oil shale could not easily be attained in the Colorado-Utah area without supplementing the surface water supply by other means such as groundwater use.

If all of the presently unused surface water available in the Four Corners region were used for synthetic fuel development, it could support at most a solvent refined coal production level equivalent in heating value to about 1×10^6 barrels/day. The production level in other fuels would range between 0.40×10^6 and 0.67×10^6 barrels/day in equivalent heating value. In the Green River Formation a 1.2×10^6 to 1.9×10^6 barrel/day industry could be supported if all of the presently unused surface water available were used for oil shale development.

We conclude by noting that we have not presented energy scenarios but only water consumption requirements for a representative level of development.

Chapter 9. References

1. Fluor Utah, Inc., "Economic System Analysis of Coal Preconversion Technology, Vol. 2, Characterization of Coal Deposits for Large-Scale Surface Mining," Report No. FE-1520-2, Energy Res. & Develop. Admin. Washington, D.C., July 1975.

2. Averitt, P., "Coal Resources of the United States, January 1, 1974," Geological Survey Bulletin No. 1412, U.S. Gov't. Printing Office, Washington, D.C., 1973.

3. National Academy of Sciences, *Rehabilitation Potential of Western Coal Lands.* Ballinger Publishing, Cambridge, Mass., 1974.

4. McKee, J. M. and Kunchal, S. K., "Energy and Water Requirements for an Oil Shale Plant Based on Paraho Processes," *Quarterly Colorado School of Mines* **71** (4), 49–64, Oct. 1976.

5. U.S. Energy Res. & Develop. Admin. and U.S. Dept. of the Interior, "Synthetic Fuels Commercialization Program—Draft Environmental Statement," Report No. ERDA-1547, U.S. Gov't Printing Office, Washington, D.C., 1975.

6. U.S. Department of the Interior, "Final Environmental Statement for the Prototype Oil Shale Leasing Program," Vol. I, U.S. Gov't. Printing Office, Washington, D.C., 1973.

7. Hendrickson, T. A., *Synthetic Fuels Data Handbook.* Cameron Engineers, Inc., Denver, Colorado, 1975.

8. Keishin, D. W., "Resource Appraisal of Oil Shale in the Green River Formation, Piceance Creek Basin, Colorado," *Quarterly Colorado School of Mines* **70** (3), 57–68 (1975).

9. U.S. Department of the Interior, "Westwide Study Report on Critical Water Problems Facing the Eleven Western States," U.S. Gov't. Printing Office, Washington, D.C., 1975.

10. Geraghty, J. J., Miller, D. W., Leeden, F. Von Der, and Troise, F. L., *Water Atlas of the United States.* Water Information Center, Port Washington, New York, 1973.

11. U.S. Geological Survey, "Water Resources Data for Montana, Part 1, Surface Water Records," U.S. Gov't. Printing Office, Washington, D.C., 1972.

12. U.S. Geological Survey, "Water Resources Data for North Dakota, Part 1, Surface Water Records," U.S. Gov't. Printing Office, Washington, D.C., 1972.

13. Geraghty and Miller, Inc., "Issues Relative to the Development and Commercialization of a Coal-Derived Synthetic Liquids Industry, Vol. III-3, Availability of Water for the Development of a Coal Liquefaction Industry," Report No. FE-1752-18, Energy Res. & Develop. Admin., Washington, D.C., May 1977.

14. U.S. Department of the Interior, "Report on Water for Energy in the Northern Great Plains Area with Emphasis on the Yellowstone River Basin," U.S. Gov't. Printing Office, Washington, D.C., 1975.

15. Northern Great Plains Resources Program, "Report of the Work Group on Water," Denver, Colorado, Dec. 1974.

16. U.S. Geological Survey, "Water Resources Data for Wyoming, Part 1, Surface Water Records," U.S. Gov't. Printing Office, Washington, D.C., 1972.

17. U.S. Geological Survey, "Water Resources Data for Arizona, Part 1, Surface Water Records," U.S. Gov't. Printing Office, Washington, D.C., 1971.

18. U.S. Geological Survey, "Water Resources Data for New Mexico, Part 1, Surface Water Records," U.S. Gov't. Printing Office, Washington, D.C., 1971.

19. U.S. Department of the Interior, "Report on Water for Energy in the Upper Colorado River Basin," U.S. Gov't. Printing Office, Washington, D.C., 1974.

20. U.S. Geological Survey, "Surface Water Supply of the United States, 1966–70, Part 9, Colorado River Basin, Vols. 1, 2," Water Supply Paper No. 2124, U.S. Gov't. Printing Office, Washington, D.C., 1973.

21. Week, J. B., Leavesley, G. H., Weider, F. A., and Saulnier, G. J., Jr., "Simulated Effects of Oil-Shale Development on the Hydrology of Piceance Basin, Colorado," Geological Survey Professional Paper 908, U.S. Gov't. Printing Office, Washington, D.C., 1974.

22. Price, D. and Arnow, T., "Summary Appraisals of the Nation's Ground-Water Resources—Upper Colorado Region," Geological Survey Professional Paper 813-c, U.S. Gov't. Printing Office, Washington, D.C., 1974.

23. Smith, W. H. and Stall, J. B., "Coal and Water Resources for Coal Conversion in Illinois," Cooperative Resources Report No. 4, Illinois State Water Survey and Illinois State Geological Survey, Urbana, Illinois, 1975.

24. U.S. Water Resources Council, "Water for Energy Self-Sufficiency," U.S. Gov't. Printing Office, Washington, D.C., 1974.

Appendix A
Conversion Factors

In the following table the conversion between weight of water and volume of water is based on a density of 62.3 pounds per cubic foot. This is the density of water at 68°F (20°C). For this density there are 8.33 pounds of water per gallon.

Multiply	By	To Obtain
acres	4.05×10^{-1}	hectares
acres	4.36×10^{4}	square feet
acres	4.05×10^{-3}	square kilometers
acres	1.56×10^{-3}	square miles
acre-feet	4.36×10^{4}	cubic feet
acre-feet	3.26×10^{5}	gallons
acre-feet/year	1.38×10^{-3}	cubic feet/second
acre-feet/year	3.91×10^{-5}	cubic meters/second
acre-feet/year	6.20×10^{-1}	gallons/minute
acre-feet/year	8.93×10^{-4}	million gallons/day
atmospheres	1.47×10	pounds/square inch
barrels, oil	4.2×10	gallons
Btu	2.52×10^{2}	calories
Btu	3.93×10^{-4}	horsepower-hours
Btu	1.06×10^{3}	Joules
Btu	2.93×10^{-4}	kilowatt-hours
cubic feet	2.30×10^{-5}	acre-feet
cubic feet	2.83×10^{-2}	cubic meters
cubic feet	7.48	gallons
cubic feet	2.83×10	liters
cubic feet of water	6.23×10	pounds of water
cubic feet/second	7.24×10^{2}	acre-feet/year
cubic feet/second	2.83×10^{-2}	cubic meters/second
cubic feet/second	4.49×10^{2}	gallons/minute
cubic feet/second	6.46×10^{-1}	million gallons/day
cubic meters/day	2.64×10^{-4}	million gallons/day
cubic meters/second	2.56×10^{4}	acre-feet/year
cubic meters/second	3.53×10	cubic feet/second
cubic meters/second	1.59×10^{4}	gallons/minute
cubic meters/second	2.28×10	million gallons/day
feet	3.05×10^{-1}	meters
gallons	3.07×10^{-6}	acre-feet

Multiply	By	To Obtain
gallons	2.38×10^{-2}	barrels, oil
gallons	1.34×10^{-1}	cubic feet
gallons ---	3.79-----------------	-liters
gallons of water	8.33	pounds of water
gallons/minute	1.61	acre-feet/year
gallons/minute	2.23×10^{-3}	cubic feet/second
gallons/minute	6.31×10^{-5}	cubic meters/second
gallons/minute -------------------------	1.44×10^{-3}--------	-million gallons/day
gallons/minute of water	5.00×10^{-1}	thousand pounds/hour of water
horsepower	6.11×10^{4}	Btu/day
horsepower	2.55×10^{3}	Btu/hour
horsepower	7.46×10^{-1}	kilowatts
horsepower ---------------------------	7.46×10^{2} ---------	-watts
inches	2.54	centimeters
kilograms	2.20	pounds
kilowatts	8.19×10^{4}	Btu/day
kilowatts	3.41×10^{3}	Btu/hour
kilowatts -----------------------------------	1.34-----------------	-horsepower
kilowatts	1×10^{3}	watts
kilowatt-hours	3.41×10^{3}	Btu
megawatts	8.19×10^{7}	Btu/day
megawatts	3.41×10^{6}	Btu/hour
megawatts ------------------------------	1×10^{6} -------------	-watts
meters	3.28	feet
miles	5.28×10^{3}	feet
miles	1.61	kilometers
milligrams/liter	1	parts/million
million gallons/day----------------------	1.12×10^{3}---------	-acre-feet/year
million gallons/day	1.55	cubic feet/second
million gallons/day	4.38×10^{-2}	cubic meters/second
million gallons/day	3.79×10^{3}	cubic meters/day
million gallons/day	6.94×10^{2}	gallons/minute
million gallons/day of water-----------	3.47×10^{2} ---------	-thousand pounds/hour of water
ounces	2.83×10	grams
pounds	4.54×10^{-1}	kilograms
pounds of water	1.20×10^{-1}	gallons of water

Multiply	By	To Obtain
pounds of water	1.60×10^{-2}	cubic feet of water
pound moles of gas----------------------	3.80×10^{2} ----------	standard cubic feet of gas
pounds/square inch	6.80×10^{-2}	atmospheres
pounds/square inch	7.03×10^{2}	kilograms/square meter
pounds/square inch	6.89×10^{3}	Newtons/square meter
square feet	2.30×10^{-5}	acres
square miles----------------------------	6.4×10^{2}-----------	acres
temperature, °C	9/5	+32 − °F
temperature, °F − 32	5/9	°C
thousand pounds/hour	1.2×10	tons/day
thousand pounds/hour	4.38×10^{3}	tons/year
thousand pounds/hour of water --------	2.00------------------	gallons/minute of water
thousand pounds/hour of water	2.88×10^{-3}	million gallons/day of water
tons	2×10^{3}	pounds
tons	9.07×10^{-1}	metric tons
tons/day	8.33×10^{-2}	thousand pounds/hour
tons/year-----------------------------	2.28×10^{-4}--------	thousand pounds/hour
watts	3.41	Btu/hour
yards	9.14×10^{-1}	meters

Appendix B
Properties of Selected Elements

Name	Symbol	Atomic Weight	Common Valence
Calcium	Ca	40.1	2+
Carbon	C	12.0	4−
Chlorine	Cl	35.5	1−
Hydrogen	H	1.0	1+
Magnesium	Mg	24.3	2+
Nitrogen	N	14.0	3−
Oxygen	O	16.0	2−
Sodium	Na	23.0	1+
Sulfur	S	32.1	2−

Index

Koppers-Totzek process, medium-Btu gas, 94,
  105–107
  ash, 196
  gasifier, 105, 106, 144
  process condensate quality, 118–121
  process water, 124
  standard plant, 106, 107
  steam for, 224

Lignite. *See* Coal
Lime, 41, 220
Lime-soda softening, 218, 220, 224
Lime softening, 218, 220, 225
Limestone, 41
Liquefied petroleum gas, 165
Liquid hydrocarbon synthesis. *See* Synthesis
Little Colorado River, 253–256
Little Missouri River, 251
Low-Btu gas, 23, 31
Low-Btu gas from coal, 23, 24, 82–84, 86. *See
  also* Lurgi process
  standard plant, 124
Lower Colorado River Basin, 254
Lower heating value, 27
Lurgi Gesellschaft, 104, 227
Lurgi process, 94, 102–104
  ash, 196
  gasifier, 87, 102
  low-Btu gas, 102
  process condensate quality, 118–120, 226
  solvent extraction for, 235
  steam for, 223
Lurgi process, medium-Btu gas, 104, 105, 179
  slagging gasifier, 104, 105
Lurgi process, standard plant, 103–104, 112–
  115. *See also* Synthetic fuels, standard plant
  ash, 197
  coal feed, 103, 104, 197
  conversion efficiency, 112–114
  cooling water, 114
  El Paso plant, 104, 113
  heat balance, 114
  hydrogen balance, 103, 104
  methanation water, 123
  North Dakota plant, 113, 114
  process water, 122–124
  product, 103
Lurgi-Ruhrgas retort, 153, 162

Marlstone, 28
Material balance. *See* Hydrogen balance
Medium-Btu gas, 23, 31, 104, 107
Medium-Btu gas from coal, 23, 24, 33, 86. *See
  also* Koppers-Totzek process, medium-Btu
  gas; Lurgi process, medium-Btu gas
  standard plant, 124

Methanation, 34
  condensate, 83
  reaction, 32, 135
  water quality, 34, 121, 124
  water quantity, 123, 124
Methanator, 33, 34
Methane
  carbon/hydrogen ratio, 14
  chemical formula, 14
  heating value, 110
Methanol
  reaction, 37
  synthesis, 23, 24, 36, 129, 131, 178
Mine-plant complex. *See* Integrated plant; Standard plant
Mining water, 5, 15, 190–195, 206–212, 265–
  267, 269. *See also* Coal mining, water; Coal
  preparation, water; Dust control, water; Oil
  shale mining, water; Oil shale preparation,
  water; Reclamation, water; Sanitary-potable
  water; Service-fire water
  quality, 215
Mississippi River, 261–263
Missouri River, 250–252
Molten Salt process
  gasifier, 88
  methanation water, 123
  process water, 122–124

Northern Great Plains
  coal resources, 246, 247
  demonstrated coal reserve, 17
  Fort Union coal region, 246
  Powder River coal region, 246
  water resources, 250–253
Northern Great Plains Research Project, 252

Occidental Oil *in situ* process, 163
Ohio River, 259–263
Ohio River Basin, 262, 276
Oil from coal. *See* Coal liquefaction
Oil from oil shale. *See* Oil shale processing
Oil shale
  composition, 29
  Fischer assay, 28
  hardness, 19
  heating value, 28, 165
  high grade, 3, 19, 28
  inorganic matter, 29
  kerogen, 18, 28, 155
  oil yield, 19, 28
  organic matter, 19, 28
  pyrolysis temperature, 18, 28, 29
  soluble bitumen, 28
  spent residue, 16, 19
  ultimate analysis, 28